A FLAWED PHYSICIAN'S TALE

First published 2025

Published under licence by Brown Dog Books and
The Self-Publishing Partnership Ltd, 10b Greenway Farm, Bath Rd,
Wick, nr. Bath BS30 5RL, UK

www.selfpublishingpartnership.co.uk

ISBN printed book: 978-1-83952-871-2
ISBN e-book: 978-1-83952-872-9

Cover design by Kevin Rylands
Internal design by Andrew Easton

Printed and bound in the UK

This book is printed on FSC® certified paper

FSC
www.fsc.org
MIX
Paper | Supporting responsible forestry
FSC® C013604

A FLAWED PHYSICIAN'S TALE

Sexual Medicine and Beyond

DR DAVID GOLDMEIER

BROWN DOG BOOKS

Contents

PREFACE

'Whether I shall turn out to be the hero of my own life, or whether that station will be held by anybody else, these pages must show'

David Copperfield by Charles Dickens

It is late autumn. I gaze at my garden from a picture bedroom window. The clematis clings to the greying, close board fencing. The lush, summer green, uncut grass, and the fig tree that has born its first fruit this year, keep my attention momentarily. But the apple tree catches and retains my attention. Its crimson fruit proudly boasted this year's harvest. The apples were just there. No one came to pick them. Maybe no one saw them – only me. The bough and branches laboured over the year, sap rising in the early spring, producing a display of white and pink early blossoms, giving way to the summer greenery. The crop fell, fully laden, and in due course will decay and fertilise the ground beneath, bringing new life with it. The branches now stand bare and stark.

I have just returned from a farewell lunch given in my honour to celebrate my retirement from clinical practice and to mark my 50 years' work in the NHS. I have been presented with an inspiring oil canvas of clouds, and numerous cards telling me how much my work has been appreciated over the years. The longest anyone has known me there is 30 years. But my story goes back way beyond that time. I sit and ponder and remember the people who were part of my life from its very inception. Many are no longer with us. I become dewy eyed and sentimental.

This book contains memories of my life. Facts. Events. Some were seemingly trivial. Others monumental and life-changing. Yet others tiptoed and hushed themselves silently into my consciousness – an odour, a taste, a remembered vista, sounds, melodies, rhythms. My tale is in the main written in chronological order. The memories stimulate other remembrances – the physician and scientist that I am unconsciously connects some memories with my research and those of others, including those of my colleagues.

I gaze out of the window again.

Sunset now covers the garden. The colours of the day fade into shades of black and grey. As I ponder, the moon rises in the early October sky, heralding the old faithful Jupiter and Saturn, unaltered and constant companions of the lengthening nights.

Many great figures in our civilisation are better understood by detailing their shortcomings and eccentricities, which may paradoxically enhance their celebrity. Winston Churchill had his alcohol, cigars and extramarital affair. Isaac Newton was seemingly a most unpleasant and jealous individual. Albert Einstein's second wife had to put up with his multiple affairs. George VI gave inspiring speeches even though he had a terrible stammer. Marie Stopes was a great believer in eugenics plus an ardent admirer of Adolf Hitler. In each case I would suggest, knowing details of their personal lives makes for a more accurate understanding of their known achievements and builds up a more colourful and complete picture of who they were and the background to their work. Expanding on this theme for example, Newton was a lonely and emotionally abandoned child, often left to fend for himself. His genius in describing the laws of gravitation and invention of the calculus necessitated him working for years on these inventions in isolation.

My own life story is more banal. And yet I feel you may understand the medical aspects of this book better by knowing more about me.

The poems, narrative accounts of factual events and illustrations were written over many years. They attempt to encapsulate in art or the written form a feeling or emotion that arose at the time.

But now I need to find a moment of clarity and inspiration to initiate my story. Past events and emotions are jumbled as I strive to remember the order of milestones and their concomitant sentiments. They do not easily condense and transmute into the written word. I wait. And wait.

Next day the garden remains inviting on this rather warm October day in the Covid era. A robin sits proudly on my 15-year-old, non-fruit bearing olive tree, which came as a cutting in a pot from the Eden Project. Will his brief red-breasted arrival help to marshal my thoughts? Too late he has flown off. The newly mown grass is too ordered to stimulate my brain. The lawn's regular chartreuse and lime stripes soothe, but do not inspire.

There is a hushed, humming roar overhead. I gaze upward. The sky is azure and cloudless. The vapour trails of two planes, one behind the other, cut through the airspace above, going from east to west – I'd guess up at just below 40,000 feet – so likely going to North America. I remember trips to conferences in the USA and happy times looking out at the vastness of the Atlantic from within a high flying plane – suspended between work and the meeting, with a tabula rasa mind.

I am ready to recall and write.

An important note. Names and places are changed in this narrative to protect the privacy and dignity of the patients described, except where they appeared in the public domain.

Chapter 1

BEGINNINGS

A four beat noise blasts and echoes into the frenzied chambers of my light sleep. What the hell is that racket? My arm seems to know, as it gropes the air, fumbles and silences my iPhone alarm. Eyelids leaden and non-responsive.

'Come on David, put in some effort.'

Suitably chided, they slowly open. I blink, and find I am staring out of the window at the naked branches of a plane tree. Fairy lights on them? No – drops of rain on the twigs reflecting the street lamp's lemony hue. My attention wanders to the black phone line that straddles the space between the wooden post office pole and my house, and I marvel that multiple online programmes can all go down that one wire and end up as distinct entities at the user end. I look back at my phone. It tells me it is 3.45 am on 14th March. I know the year is 2020. Why am I waking so early? My mind clears a little. That's right – I lead the Jane Wadsworth Sexual Function Clinic at St Mary's Hospital London. I am considered a world expert in sexual medicine, and I rise well before dawn to get to work early. But something's not right. I try to sit up. It takes all my strength. A Herculean effort. Core muscles and limbs moving like they were immersed in thick treacle. My legs are dead weights. I summon up resolve and push them over the edge of the bed. I stand erect, but immediately the world spins, and I find myself immersed in a grey swirling mist. I think I am going to faint, so I flop down on the bed again. I take a sip of water from the bedside table. It makes me cough. Not an 'ahem, or tickle at the back of the throat' once-off cough, but a succession of involuntary explosions, ending with me feeling utterly drained and bringing up a large glob of yellow sputum. I notice I am wet all over and realise I am sweating profusely.

As I try to work out what's going on, I am speared in the solar plexus. Fear injects a throbbing remembered thought into my now clearing mind. One of my junior doctors went down with coronavirus a few days ago. Maybe I am infected too? And Tricia my wife, how is she? Last night she was confused. Mumbled nonsense as she wrung her hands, as if her palms could in some way squeeze sense out of gibberish. My brother almost certainly has coronavirus. He is seriously ill, fighting for his life in the local district hospital. The latest I heard is that his oxygen levels were low. They turned him over on his front – seemingly that helps optimise lung function. I find out later that day that he is going downhill. Nowadays 73 is young. But he has a nasty skin disease for which he takes immunosuppressants. So he is vulnerable, highly vulnerable.

The panic has injected enough energy into me, so I finally stand up. Steady now, take it slowly. I walk a few steps towards Tricia's bedroom. Two steps and the world spins. My vision is hazy again. I fall down in a faint. The overflowing washing basket catches me mid descent. My horizontal pose sends blood back to my brain. It raises my blood pressure. Panic returns. Tricia cries out 'What's that?' I stumble into her room. She's curled up in a ball that looks so small in that king-size bed. I ask her how she is. She tells me she is in pain. I ask if she has taken her oxycodone tablets.

'What are they?' she mumbles.

She is obviously still muddled. Doesn't know where she is or the day of the week. She has a temperature but no cough. I give her her morning medication, which she swallows obediently but refuses her usual morning coffee.

I sit on her bedside and try to make sense of my situation.

I am ill.

Tricia is ill.

Michael is dying.

It is five hours till the GP surgery opens.

Do I take Tricia to hospital?

Am I well enough to drive?

I know that the outside world is in lockdown.

I decide to wait till I have spoken to the GP. I am left alone with my uninvited companion – anxiety that pulsates and throbs and delivers its poison into mind and body.

I wonder if Tricia and I are dying.

'Maybe,' I reply to myself.

My mind is now in freefall, panic leading me to where I do not want to go.

An Einsteinian quote crashes into my consciousness: *'Our situation on this earth seems strange. Every one of us appears here involuntarily and uninvited for a short stay, without knowing the whys and wherefores.'*

Is my short stay at an end?

Shakespeare follows in quick succession.

'We are such stuff as dreams are made on, And our little life is rounded with a sleep.'

I shed a tear. Is my ' little life' drawing to a close?

Terror jerks me back to reality.

Naked fear then searches out and affixes itself onto painful memories. It presents them to me unbidden, as I lie on my bed, over-breathing, sweating, shaking.

Fear of being abandoned in hospital as a two year old. Crying for my mother.

Fear of speaking in class. Can't say my name. The words just won't come out. D … d ... d. I should give up. I am so ashamed. I'm such an idiot. I wish I could disappear. I try again and contort my face and screw up my eyes. I spit saliva as my 'd.d.d.d' continues. But no intelligible word is formed. My repetition has now changed to a block. I so want to push the word out. All my breath pushes hard against my closed lips. I feel hot. I think my face must be red. The teacher grabs my now downcast head and firmly looks me in the eye. 'Try harder David, put some effort into it'! I am only five, as are the girls whose giggling bursts into overt laughter as they are daily entertained by my efforts to enunciate.

Wave upon wave of panic in a locked psychiatric ward, where no one seems to care.

I fall into a nightmarish sleep, as black and thick as the darkest smog filled night, pierced by flashes of terror.

As I lie there, the virus, already embedded in my nose, has unbeknown to me, travelled up my olfactory nerve to reach my brain.

Come 9 am I try to speak to the GP. The recorded message says no one should attend the surgery until invited to by the practice. I am on the phone for two hours but cannot get through. I give up.

A week later Tricia is better. Michael is dead.

Two weeks later my cough disappears. My fear is no better.

I finally contact the GP who tells me to take Vitamin D.

I live in a state of terror for 16 months.

I then fall into the hands of a wonderful NHS psychiatrist. She and my eldest son encourage me to take the correct antidepressants. After six weeks of exacerbated terror and vivid nightmares the medication finally works – a small haven of peace. But the territory I find myself in is a novel and strange land.

To make sense of where I find myself now I must look at how I arrived at this juncture in my life. I ponder and wonder. Then write.

And so my tale unfolds.

Chapter 2

FAMILY BACKGROUND

My father's family hail from Kleinlangheim, a small village in Lower Franconia, which is just over a mile from the upper waters of the River Main, as they meander slowly through one of the best wine producing areas of Bavaria. Justin, my father, taught himself to swim in the Main, accompanied by river rats. He told me he naturally gravitated to sidestroke, as it did not cause the headaches of keeping the head above the water as in breaststroke. Jews had resided in Kleinlangheim since 1415, although the first record of Jews in Germany dates from the end of the Roman occupation in the mid-4th century of the Common era. The Jewish

The Goldmeier farmhouse in Kleinlangheim, as it appears today

population of Kleinlangheim was 118 in 1933 and 38 in 1937. The Goldmeier family, like most of their neighbours, owned a small farm consisting of a minute strip of arable land plus one cow. Marriage documents described them as 'cattle dealers'. In any event they were very poor. I can imagine the family eating their frugal meals off an aged wooden table. Bread, boiled root vegetables, water and coffee during the week, with possibly some meat on the Sabbath. Milk if the cow was a good buy. During the Depression in the early 1930s wheelbarrows full of millions of Deutsche Mark notes were exchanged for a loaf of bread or a pound of butter. Woodburning stoves for heat in the winter. Candlelight to read books by in the evenings. As the eldest sibling, my grandfather Louis 'inherited' the farm and the farmhouse.

In 1813, Bavaria passed a 'Jew Edict' law, which among other conditions stipulated that every Jew was required to have a surname. Meier may mean 'major' in German, or 'from light' or 'sparkling' in Hebrew. Extrapolating to the possible meaning of the name Goldmeier makes little sense, as the family had no gold, nor were they sparkling or indeed in any way important. The irony may have been deliberate! The edict also set out that Jews were to be taught to speak 'Modern German' rather than the old German Jewish dialect, had to send their children to public schools and lastly that only the eldest son in a family was allowed to marry. My grandfather's male siblings therefore had to move out of Bavaria. One of his brothers, Jacob, opened what became a very successful retail wine business in Frankfurt. Justin, was born in 1909 as, an older child he walked a 12-mile round journey to and from secondary school in Kitzingen each day, where one of his classmates was Rainer Kesselring, adopted son of the infamous World War 2 Luftwaffe Field Marshal Albert Kesselring. My father informed me that a 'state of the art' fighter plane had landed in Kleinlangheim in 1916. The whole village went out to a grass field to see the monoplane which was probably the Fokker Eindecker, the highly successful German fighter that was the scourge of the allied aircraft, as the propellers' gears were synchronised with its machine guns, and so could be fired through the propeller by the single airman. My father was taught some science in his early teens and was told by his prescient, and by all accounts well-read teacher, that Albert Einstein's famous equation, $E=Mc^2$ (circa

1905 in the Special Theory of Relativity) implied utter destruction if ever people learnt to harness the energy in matter! My father left school aged 16, in effect to become the family breadwinner, working for his Uncle Jacob in Frankfurt, selling wine in a number of European countries, including France, England, Scotland and Ireland. He quickly became a proficient English and French speaker while learning on the job. In his travels to the UK, he made a number of important contacts, so that he was able to bring his father, mother, brother, two cousins and Uncle Isidore to safety in England in the late 1930s. The British government at that time required a guarantee of £50 (about £2,000 in today's value) for each person who had no employment, so that they would not become a financial burden to the country. In most instances either the British Jewish community or the Quakers funded this. The fee, however, could be waived if the immigrant was a professional or was in 'domestic service'. How exactly my father found the money to pay for his relatives, I do not know. Maybe he had savings or asked others for help. Uncle Jacob from Frankfurt along with 15 other Goldmeiers perished in the Holocaust in World War 2. Jewish Kleinlangheim was effectively erased. Its synagogue and scrolls of the law were torched on 'Kristallnacht' in 1938, and its small primary school building demolished. The Kleinlangheim Jews who did not emigrate were all later exterminated in concentration camps.

In late 1939 my father was 'interned' in the Isle of Man as an 'enemy alien', along with thousands of other men, women and even some children, most of whom came from Germany and Austria. Refugees were arrested in their homes, workplaces and offices, and shovelled into police vans, often to the sounds of the local community booing and hissing them. No explanations were given by the police, 'Just get your clothes on and come with us'. Among the internees were a handful of Nazi sympathisers. Since a German-speaking Jew could not easily be differentiated from a Nazi, Churchill had issued his famous edict 'Collar the lot'. My father was 'imprisoned' in the local five-storeyed Edwardian terraced boarding houses by the seafront in the capital Douglas, which until the start of the war had been filled with holidaymakers from mainland UK. The hotel owners were displaced to live elsewhere on the island. The internees were surprisingly well fed on local seafood and dairy produce, and had a reasonable degree of freedom within the barbed wired fences that surrounded the different prison areas. Jewish internees came from a wide strata of society, and included the religiously orthodox, atheistic, lawyers, musicians and other professionals. A German Jewish professor of law at Cambridge was arrested and hustled into a Black Mariah police van before he had time to unlock the safe wherein were university exam papers. Because of that the finals exams had to be cancelled. Some 200 Jewish doctors helped the local practitioners run sick bays for the internees. Renowned artists would sketch using a mixture of margarine and graphite made by burning twigs. Sculptors made wooden reliefs from local discarded wood. Guards would often turn a blind eye to inmates who went out of the barbed wire enclaves to buy some food at local shops. There were at least 56 deaths among the detainees, many of which were suicides. They were buried in the local municipal cemetery, the cost of transporting the bodies to the mainland was deemed too expensive. Finding details of those who passed away on the island at that time remains under the jurisdiction of the Manx coroner, who even now does not make it easy to obtain such information. Most Jews were released within a few months after being interviewed and interrogated. I am unsure of exactly what my father did once released from the island six months later, but I do know he became an air raid warden in the Hampstead area during the worst of the Blitz. He was meant to patrol with another German Jew. However that man was too frightened of the bombing, so my father walked the Blitzed streets alone. Towards the end of the war he started up a pickled food business in Stepney Green in London's East End – I presume this was largely after the bombing ended. Although my father spoke grammatically correct English he retained his strong German accent for life. That said, he insisted he was one notch above Georg Ludwig, the erstwhile Elector of Hanover, who became George I of Great Britain, who spoke very little English at the time of his ascension to the British throne.

And how do I remember my father in my early years? There are old photographs of him in baggy trousers which extended from his feet up to and beyond his umbilicus, loose fitting shirt tucked in, with dark, straight hair that he backcombed. Even though he was in his late forties when these pictures were taken, his head is slightly tilted down, giving him an appearance of a shy teenager.

Neither in those early photos nor in person (they call it 'real time' now), did he seem to have retained any of the traumas or vicissitudes of his previous life. The little finger on one of his hands, I can't recall which, was permanently flexed, likely due to Dupuytren's contracture, where there is a fibrous overgrowth in the tendon sheaths of that finger. He cleaned his nails every morning in the bathroom. The detritus he had picked up the previous day was then transferred onto a piece of toilet paper and disposed of in the nearby lavatory. Of course toilet paper in those days, we used the brand called Izal, was shiny, hard and medicated. Even in the mid 1950s, people in many parts of the UK were distrustful of soft toilet paper because they were not 'reliable'. Back to my father. He seemed to have an almost inexhaustible supply of jokes, which he seemed to find funnier than his listeners. Maybe it was the accent. He explained to me that 'An Englishman laughs twice at jokes, once out of politeness, even though he does not understand it, and then a few minutes later when the penny drops'. We did not have a close physical relationship but when I was out with him he would hold my hand, which truth be known, I found a trifle disagreeable. In his early 50s he learned to drive. After he passed his driving test he took all four of us for a ride in the country in our newly purchased off white Austin A50. He nearly killed us on that first outing by being far too slow in overtaking a lorry. We nearly smashed head on into an oncoming car. A large part of the problem was that the average cars of those days were heavy and underpowered. And on that occasion there was also the weight of four passengers.

I never saw my father smoke. However, he had previously smoked heavily, telling me that in his heyday he only lit up one cigarette a day. The ensuing 59 would be ignited on the stub of the last one. A doctor in the late 1940s had strongly advised him to quit smoking. This advice was of course quite correct but ran in the face of the medical profession's at best neutral attitude to smoking at the time. In 1946 Reynolds launched an ad campaign with the slogan 'More doctors smoke Camels than any other cigarette'. It was only in 1957 that Doll and Hill published their paper showing the devastating effects of smoking on health. Seemingly my father had to replace this oral gratification with something. So for most of his later life he sucked and bit on an empty pipe, almost always holding it upside down in his mouth.

As he grew older he put on quite a middle-aged spread. At no stage in his life did he consider dieting or exercising.

My mother, Helen Pick's early life was more mysterious. She was born in 1908 in or near what was then Breslau in eastern Germany's Upper Silesia – but now renamed Wroclaw in Poland. Breslau had a Jewish presence since the 13th century. In the 19th century Orthodox and Liberal Judaism flourished together. Robert Koch, the founder of modern bacteriology was raised in the city. In the mid 1800s a Jewish religious seminary and orphanage were set up. In 1925 the Jewish population of Breslau was 23,240. Most of the community who did not emigrate succumbed in the Holocaust.

My mother lost her mother to puerperal fever when she was five. Her and her three siblings were placed in a local Jewish orphanage about the time of the onset of World War 1. My mother told me that the four siblings wanted to stay in each other's company as much as possible. A sound proposition looked at from the 21st century. However the orphanage would have none of it, and separated them within 24 hours of arrival. Those years without a parent or sibling to touch or to hold must have wreaked psychological havoc in those young minds. Why was it done? German orderliness, so that children were sorted by age? Possibly. My mother told me she did not know. Maybe she knew but didn't want to discuss it. Just too painful. Maybe best left unspoken. But it did leave a great void in the details of those years. You may ask, why did their father Martin not look after them? Martin volunteered or was conscripted into the German Army at about the same time as they were abandoned as orphans. Martin won the Iron Cross Medal for bravery, of which 230,000 were awarded during the war. I attempted to discover more about Martin's war record by contacting German military records in Bremen. I was told that all World War 1 records had been destroyed in World War 2 during Allied bombing of the city. It is possible Martin remarried before going off to war. His second wife bore him a number of children.

My mother reappears in records in the 1930s when she worked in a Jewish respite centre in Lehnitz near Berlin as a nurse-cum-domestic supervisor. The Lehnitz centre functioned during the Nazi period and had allowed Jews, particularly young people, to escape the stress of anti-Semitism in and around

Berlin, for rest, recuperation and inspiration. In 1938 she and four other young women in Lehnitz were approached by Dr Bernard Schlesinger, a consultant paediatrician at Great Ormond Street London to help recruit and organise the transfer of themselves and 12 'educated' children and teenagers from Berlin to the UK. A word About Dr Schlesinger. His father Richard had emigrated to the UK in the 1880s from Frankfurt because he felt more money could be made on the London Stock Exchange compared to the one in Frankfurt. He became a strong Anglophile and almost worshiped the British royal family. Bernard served in the Royal Army Medical Corps in both World Wars and reached the rank of brigadier. One of Bernard's sons was John Schlesinger, who became a famous film director, winning awards for *Midnight Cowboy* and *Far from the Madding Crowd.*

The transfer of these 12 children was part of the greater 'Kindertransport', which allowed 10,000 children from Nazi occupation to find shelter in the UK, albeit mainly without their parents. For Dr Schlesinger it necessitated quite a deal of arranging and planning. He also provided the finances needed to support the children in a large rented house in Shepherds Hill Highgate, where they all stayed for 9 months until the general pre-Blitz evacuation. At some point prior to my mother coming to the UK, she found out her brother Arthur had been arrested by the Gestapo and thrown into Buchenwald concentration camp. Somehow she made the 380 mile round trip and persuaded the Nazis to release him. Ever since I first heard this harrowing tale I have asked myself, 'why should the Nazis have released him?' I conjecture. My mind wanders to multiple possible scenarios. But I never did ask. Likely my mother would have not told me the whole truth anyway. Arthur was able to find a berth on one of the last boats to leave Germany, eventual destination Shanghai, where he spent the war years, finally settling in the USA and bringing up a family. My mother's older sister Elsa died on the way to Auschwitz, having moved temporarily to safety in Holland in 1938. Prior to this she met my mother who was on her way to the UK with 'her children', and tried to persuade her to stay in 'safe' Holland.

The train lurched to a halt, throwing the children from their seats. Matron Helen registered their pallor and taciturnity as they slowly clambered seatwards. They looked scared. Petrified in fact. She knew each child had a soul, but sensed each child's spiritual self had been subdued into submission by the Nazi terror and forceful separation from their parents.

Helen's task was to accompany them from Berlin, via Holland to safety in England. A different obligation would make that task challenging. Compassion and responsibility would vie with love.

They had arrived at the Hook of Holland, ready to transfer to a cross channel ferry. She helped the children off the train. Each wore a stringed, wooden neck nametag and carried a small suitcase containing one item. Meir took his favourite teddy bear. Hansel her best doll.

Not 24 hours before, Helen witnessed the children's parents, some weeping with arms outstretched, others bravely holding back tears, but all wanting safety for their child. Wishing them gone and godspeed. Every parent sitting with the pain of loss.

The children were distressed and confused. Yet in spite of their distress, Helen had heard them whispering about England, the King, the young princesses and soldiers with furry hats.

Helen gathered her six wards at the foot of the train. Some clung to her drab brown overcoat. Others grabbed a hand or an arm. A dark drizzle fell. Helen too felt panicky. Was she here? Was that them? Yes!! Helen was overjoyed – Else and Fritz were here! Else surged towards her. Else, who had soothed Helen as a toddler in the orphanage on the rare occasions they met. A mother figure in dark times. And yet, Helen now had her own responsibilities. She was, anchored by the children, standing motionless as Else fell on her. The sisters wept. The children became entangled in the outpouring of emotion.

Presently Else handed each child a plastic cup of hot soup, a sandwich and a bar of chocolate. Their hunger overcame their misery.

Helen knew what was coming next. Else implored her to stay in Holland, which she assured Helen was Nazi free. Helen wanted to believe her. But she had given her word to deliver these children to London and Dr Schlesinger's hostel.

'Ich kann es Nicht'.

The sisters embraced for a final time. With a broken heart Helen and the children boarded the ferry. No one looked back.

In years to come most of the children would find out that they were the only survivors of their families.

My mother's younger sister Betty made her way to the USA. Their father Martin perished in a cattle truck on the way to Auschwitz concentration camp in 1942.

My mother had a trim figure all of her life. Her light brown hair was thin and fine in texture. Her sepia wedding picture shows her with her head cocked to one side, smiling broadly and generously. To me the pose speaks of good heartedness, and a willingness to be kind to everyone. Helen, who was addressed as Helene by some, seemed to have had a cultured background. When happy, she sang sections of Wagner's lighter operas as she busied herself with household chores – or looked after us children – hummed Franz Lehar's *Gold and Silver Waltz*, or quietly crooned Schubert's lieder. *Die Forelle* was by far her favourite. There was always a clean smell about her. She spoke with a slight hesitancy, which may have been a mild stammer. Helen's taste in clothes was conservative, telling me that the height of fashion was to dress in black. Table manners were important to her – so no toilet when eating meals ('only the pigs do that'). And elbows NEVER on the table!

I am unclear when or where my parents met. They were married in 1945 and initially lived in Langland Gardens in Hampstead. As discussed above, my mother was by far the more cultured of the couple. She told me later in her life that early in the marriage it became obvious that my father had not bathed for some time. I suppose having a bath in Kleinlangheim was an infrequent event. She told him in no uncertain terms he needed to run a bath. He shut the bathroom door, but reappeared a couple of minutes later. She asked him what exactly he had done in the bathroom. He told her that he had merely sat in the tepid water for a minute, poured water over his head and then dried himself. Helen had to explain to him exactly how to soap and wash himself. When retelling this story she remarked, 'He was so primitive'. My father's bathroom behaviour should be seen in the context of his social class and the times he was brought up in. Bathing in a peasant house would have been in a metal tub, and the occasion of using tepid water for cleaning undertaken no more often than once a week. The whole family would likely have used the same water. Rural Germany in the early 1900s was in many ways still in the 19th century where hot baths were considered to be debilitating. Soap was used to clean the hair only. Its use to clean other parts of the body was felt by some to lead to insanity.

My brother Michael was born in November 1946 at Queen Charlotte's Hospital in Hammersmith London.

Chapter 3

BIRTH AND CHILDHOOD

I was born on 23 January 1948 at the Bearsted Memorial Hospital in Stoke Newington London N16. I am told I weighed about 8 pounds and apparently came into this world by normal delivery. A brief history of the Bearsted below suggests it functioned only for a few short years.

The Bearsted Memorial Hospital in north London

In 1937 plans began to build a new hospital to replace the small cramped Jewish Maternity Hospital in Whitechapel. A site in Stoke Newington was favoured, as by this time a large number of the younger Jewish population had moved away from the East End to north London. Considerable amounts of money had been raised for the project from pennies subscribed by poor Jews in the East End as well as from many wealthy donors, including Viscount Bearsted, after whose family the Hospital was named. Building work began in 1939 but almost immediately had to be deferred because of the outbreak of World War 2. Additional premises were acquired at Hampton Court. In 1945 permission was granted for part of the new Hospital to be built in Lordship Road N16. This first section containing 30 beds was finally completed in August 1947 and officially opened in December by Viscountess Bearsted, the president of the Hospital. The Hospital joined the NHS in 1948 and, while still mainly catering for orthodox Jewish mothers (some 80 per cent), also admitted women of other faiths. By 1969 the Royal College of Obstetricians withdrew recognition of the Hospital for postgraduate medical training because of the lack of sufficient beds which also threatened withdrawal of the midwifery training school by the Central Midwives Board, who believed that a minimum of 50 beds was necessary for satisfactory training. By 1971 the decreasing Jewish population of the area and subsequent reduction in the number of Jewish mothers admitted to the Hospital affected the training of mohelim, who performed ritual circumcisions there (17 were performed in 1971, compared to 33 in 1970 and 37 in 1969). In November 1974 the Enfield and Haringey Area Health Authority ordered the Hospital to be closed, ostensibly because of staff shortages. In spite of protests over the closure of this unique Jewish maternity hospital by the Board of Deputies of British Jews, the Chief Rabbi and the Jewish Chronicle, as well as a petition signed by over 2,000 people, the hospital was forced to close. The Department of Health and Social Security handed the Bearsted Memorial Hospital back to the Jewish community for 'social service purposes'. There is currently a Norwood family centre on that site – a purpose-built new building used for family therapies for the wider Jewish community.

My bris (Jewish ritual circumcision) I believe was on time at eight days old and carried out by Dr Jacob Snowman who also circumcised Prince Charles in the same year.

A third of men in the world are circumcised, with very high rates being reported in the USA. The practice is at least 5,000 years old, and was undertaken in ancient

Egypt. Columbus's crew in 1492 reported that some natives of the New World were circumcised. Ritual male circumcision is also common in Africa and the South Sea Islands. Judaism and Islam requirement for male circumcision is based on God's commandment to Abraham to circumcise his son Isaac. A Jewish male infant has his circumcision at eight days old. However, if the child is ill the ceremony is deferred. Critics have construed the circumcision of a small baby as a physical and psychological assault on a person who has not given consent for the operation. Evidence for the effect of the operation on the neonate, apart from crying, which often takes place as his nappy is removed, includes an increase in heart rate and a high level of the stress hormone cortisol at the time of the event. Jews reply to this by saying that bris is an emblem of religious faith and national identity, which is imprinted in their collective memory. Among the critics of circumcision was Turnus Rufus, the Roman governor of Judea in the 2nd century CE. He asked one of the greatest Jewish sages of the time, Rabbi Akiva, why the Jewish God did not form babies without foreskins. Akiva answered that in the same way as wheat is produced by God and then fashioned by man into bread, so it is for the penis. Persecution of Jews has always involved attempts at banning circumcision. In 168 BCE the Greek Syrian Antiochus Epiphanes ordered execution of all circumcised babies in Jewish Palestine. Similar persecution was undertaken by the Roman emperor Hadrian, in the Spanish Inquisition (even though Jesus himself had been circumcised) right up to the Nazi period. In the UK, highly trained circumcisers, or Mohelim, are accredited by the Initiation Society, which was founded in 1745. Minor complications affect less that 0.3 per cent of infants.

Three recent large randomised trials in Africa have demonstrated that circumcision decreases acquisition of both HIV, and genital herpes and is recommended by the World Health Organization to be undertaken in areas where the HIV prevalence is high. There is no evidence that the practices diminishes sexual pleasure.

Goodman J. Jewish circumcision: an alternative perspective. BJU International 1999 Jan;83 Suppl 1:22–7.

Tobian, AA, Kacker, S , Quinn, TC. Male circumcision: a globally relevant but under-utilized method for the prevention of HIV and other sexually transmitted infections. Annual Review of Medicine 2014; 65:293–306.

Morris, BJ and Krieger, JN. The Contrasting Evidence Concerning the Effect of Male Circumcision on Sexual Function, Sensation, and Pleasure: A Systematic Review. The Journal of Sexual Medicine. 2020 Dec; 8(4):577–598.

My family moved to 22 Pennine Drive NW2 on the Golders Green Estate just prior to my birth. The Estate was in fact nearer Cricklewood than Golders Green.

The Golders Green Estate with Handley Page's factory towards the top right of the picture © Historic England Archive (Aerofilms Collection)

The Golders Green Estate (see picture) was built on the remains of Handley Page airfield. Handley Page built and flew planes from fields near this factory up until 1929, in particular flights to Paris from 1920 onwards. The housing estate was built on the airfield in the early 1930s but the factory remained till the 1964. Handley Page continued to manufacture planes till 1970, including the famous World War 2 Halifax bomber and the Victor jet bomber which was part of the

UK's nuclear deterrent. The estate was at one time home to the Hollywood actress Jean Simmons. All the houses were semi-detached, with two downstairs rooms and a tiny kitchen.

I have a vague memory of having my adenoids taken out at Paddington Green Children's hospital. I think I must have been about two. I was not told I would have to stay for 10 days without seeing my parents. I cried a lot. I remember being very frightened. Paddington Green no longer functions as a hospital, however the building itself is listed, so still stands to this day.

A year later, in 1952, when I was three, I had my tonsils taken out in the now demolished St Andrew's hospital in Dollis Hill. The Sisters of Mercy ran the nursing aspects of treatment.

David's tonsillectomy story.

My Mummy told me l that I would be going to hospital.

'Tomorrow David.'

I ask why I am going into hospital. She says there are nasty bits of me called tonsils which must be taken away. I asked if she will stay with me in the hospital. 'Of course,' she retorted. I love my Mummy and she is kind to me. She cuddles and kisses me when I fall over or when I have a bad dream. She makes me lovely food. She is in the bedroom just next door when I wake up. I like going into her bed in the morning. It is warm. I feel happy there.

The next morning she helps me put my clothes on. There is a small suitcase on the bed. I ask who that is for? She tells me it is for me for my hospital visit. 'But can I come back with you this evening Mummy?'

'Of course David,' she says.

My Mummy never tells lies.

So she and I take a bus and then another bus. Mummy says the hospital is called St Andrew's in Dollis Hill. She and I and the suitcase walk up the gravel path. I don't like it inside. There are no nurses there, just women with large white flowing hats. Mummy says they are called nuns. We go up the stairs and into a big room with lots of cots and beds. It makes me frightened. I cling to Mummy – 'You will stay Mummy won't you?'. Mummy doesn't answer. A large nun tells me to take my clothes off and get into the bed. She is a big lady with a shouty voice like a man's and speaks like someone who is not from London. 'Now be a brave boy David'. I am crying more and hold my Mummy's leg. The nun wrenches me away and says, 'say goodbye to your mummy now David'. 'But Mummy said she was staying'.

I see my mummy leave me. She is crying too. 'Come back Mummy. Don't leave me.'

But she has gone.

'When can I go home?' I ask the nun

'After your operation.'

'What is an operation?'

'Now don't ask so many questions and stop that crying immediately.'

The nun looks angry.

I can't stop crying.

She drags me and hoists me onto the bed. 'Now stay there and don't move.'

'I want my mummy.'

I cry and cry. No one listens.

Other children are asleep or crying. Some are just staring into space. A boy near me is cuddling a large soft pussy cat toy. His eyes are all red.

I think, 'Why would Mummy leave me in this horrible place?'

The nun is angry now. 'Stop that crying or I will smack you.'

I cry more, 'I want my mummy.'

'If you don't stop crying I will lock you in the bathroom. You are a bad little boy.'

I don't know what she means.

She drags me like a sack of potatoes into a huge steamy room with tiles going up to a very high ceiling. There are three baths there. She slams the door shut.

'Now stop that stupid crying or I shall never let you out!'

I am there for so long. I stop crying. I am wondering why my mummy lied to me. I feel exhausted.

I am back on the bed and fall asleep.

When I wake up the nasty nun is there. She tells me to eat the jam on the spoon. 'I don't like jam.'

'You are a naughty boy. Eat it or I will make you eat it.'

It tastes horrid.

Soon I am asleep.

When I wake up I have a horrible pain in my mouth. I cry more for my mummy.

The nasty nun reappears. 'If you cry like that you will bleed to death. Now shut up you horrid child.'

I decide I am going home.

I walk out of the big room, down some stairs and out of the front door. But a man asks, 'Where are you going?' I say, 'Home'. He laughs. He looks kind. 'Come on young man, back to the ward with you.'

I stay in the big room for days and days. I speak to my mummy on the phone. She tells me she loves me and that I will be back home soon. 'But I want to come home now Mummy'. 'Soon,' she says.

I am waiting. They give me ice cream twice a day. I am not hungry. I have stopped crying ... I have no feelings anymore inside me. When my mummy phones again I cannot talk to her. She says, 'Hello David'. I cannot answer her. I feel cold and nothing inside. I sit and stare and hold the sheets. It must be a bad dream. I cannot get out.

More days of no feeling and no eating.

Then my mummy comes with the same suitcase. 'You are coming home now David'. I am so happy but also so angry. We go home on the buses. She puts me to bed.

I am happy to be home.

But I am worried.

'Will you be there in the morning Mummy? You won't take me to the hospital anymore?'

'Of course not David.'

My mummy never lies…

Tonsillectomy was fashionable in the pre-antibiotic era. Well conducted trials in children suggested there was a decrease in subsequent tonsillitis after this operation over a three-year period compared to those who did not have the operation. However the rate of readmission after the operation was as high as about 10 per cent. Crying after tonsillectomy increases the risk of severe haemorrhage. Death from this scenario was rare but well documented. Nowadays the operation is not so commonly undertaken, and parents are encouraged to remain with the child over a much shorter inpatient stay.

In 1952 the social worker and psychoanalyst James Robertson made a short film, A Two-Year-Old Goes to Hospital, *to demonstrate a child's grief at being separated from her parents, particularly from the mother. Initial protest (calling and crying for mother), followed by despair (quiet withdrawal) and finally detachment (hardly responding to mother when she does visit) is vividly displayed by Laura, the two-year-old subject of the film. When the film was shown in the early 1950s, the medical and nursing professions reacted with hostility to Robertson's idea that a young child should not be separated from its parents, angrily retorting that the despair and detachment phases were merely signs of the child 'settling in'. However, by 1957 professional and public concern were sufficient to cause the Minister of Health to appoint Sir Harry Platt and his committee to report on the issue of young children in hospital. The Platt committee's 1959 report 'The Welfare of Children in Hospital' promoted unrestricted visiting for parents of children of all ages in British hospitals. But it came too late for me.*

Robertson, J and Robertson, J. Separation and the Very Young. Free Association Books, London, 1989.

My education began, aged five, at Clitterhouse infants school which was a half a mile from our home. The building was erected in the 1930s. My mother took me to school each day until I was about eight years old. The building and the school still exist but has changed its name to Claremont. I was frightened at school, worrying that my mother would forget to fetch me. After all, how could I really trust her? When she was late my worst fears were confirmed.

The classrooms smelled of paints, glue and aging wooden desks. One of the pupils in my year returned after suffering severe polio, wearing callipers on both legs. These were attached to his legs with the aid of leather straps which had a certain thick musty odour. Registration time at the beginning of the day greatly distressed me. We each had to call our name out in alphabetical order. My stammer, which at times was severe, displayed blocks, repetitions and introductory words (see below), had already manifested long before school entry. Most stammerers find it very difficult to say their name (phone calls can also be difficult). The coven of the three witch-like, giggling girls waited for their daily entertainment as I tried to spit the words out. These tormentors visited me in my dreams and in my fearful thoughts as I walked to school. Their suppressed sniggering, which turned into overt raucous laughter, made me want to disappear. I felt so ashamed of myself.

Stammering, or stuttering, as it is named in North America, refers to developmental stammering rather than a speech defect coming on in older age, for example after a stroke. The word in German is stottern, in French balbutier and Italian balbettare, the latter two deriving from the Latin balbus meaning fumbling. What is a stammer? Once heard and seen it cannot be forgotten. A severe stammerer will display some or all of the following features – disordered articulation, grimacing and contortion of the facial muscles (sometimes with involuntary limb movement), use of 'starter words' (mine is 'you know' which used to annoy many listeners), blocks, where no sound at all is emitted in spite of great effort, repetition (e.g. my, my, my, my thoughts), prolongation of the word or syllable e.g.(m ... m ...

m ... my). Attempts at speech, and even anticipation of speech, come with feelings of anxiety, ridicule and difficulty in making friends, particularly in adolescence, leading to a low quality of life. The listener also feels awkward and may avoid or keep conversation to a minimum. Some listeners will enunciate the word that they guess the speaker is trying to say, adding to the fear and annoyance of the stammerer. Many of the greatest in history stammered or were reputed to have a stammer – Moses, the Emperor Claudius, Isaac Newton, George VI (the queen's father as portrayed in 'The King's Speech') and even Churchill.

Research into the cause of stammering has revealed no easy overarching explanation. However it has been shown to be associated with chromosomal abnormalities and alterations of a number of gene groups. At the brain level, scanning has shown changes in the parts of the brain involved in hearing, speech and the basal ganglia – that is the area beneath the cortex. Whether these are result of or predate speech is somewhat unclear. The brain neurotransmitter dopamine is dysregulated in stammerers. Up to 10 per cent of any population will develop a stammer, which can range in severity from a mild dysfluency up to a severe problem. Boys and men are four times as likely as girls to develop a stammer. Of those children who do develop stammers, over 90 per cent will recover with or without treatment. Treatment over the ages has included speaking with a mouthful of marbles or 'hitting the stammerer in the face when the weather is cloudy'. Nowadays the cornerstone of treatment is early speech therapy, which may include the use of music, metronomes and voluntary stammering. Devices that aid stammerers include delayed feedback of their own voice and alteration of the pitch of the person's speech. Antipsychotic drugs which work on the dopamine system such as olanzapine may be helpful to some. However they often cause weight gain. In the UK 'STAMMA' is the British Stammering Association's national charity for people who stammer and those that support them.

Perez, HR and Stoeckle, JH. Stuttering: Clinical and research update. Canadian Family Physician 2016 Jun; 62(6):479–84. PMID: 27303004; PMCID: PMC4907555.

Two important events occurred almost simultaneously in June 1953. The new, young Queen Elizabeth was to have her coronation. Our house was decorated with red, white and blue bunting. I received a mug decorated with a smiling, coloured image of Her Majesty. To mark this important event my father bought a television set. It was as much a piece of furniture as a viewing box. The screen was no larger than 12 inches square, and its 450 lines gave a fuzzy black and white picture of the event. The screen was mounted in a dark wooden cabinet.

Although my father paid the equivalent of £800 for the TV, he could not afford a sofa for all the neighbours to sit on to view the event. That being so, a bed was carefully manoeuvred from upstairs to a downstairs room and covered with rugs for the audience to use as a sofa. My memories of that day are as grainy as the TV image that was projected by the cathode ray tube onto the screen.

Aged eight I transferred to the junior school which was housed in a new, purpose-built structure. The Conservative MP Sir Frank Markham discussed the architecture of such new schools, erected in the late 1950s and early 1960s, describing them as 'shoes boxes with glass windows' and 'awful and ugly', comparing them to army huts erected during the previous war. Such was my junior school, and indeed the building had no beauty. Because of the expanses of glass it overheated in the summer and was cold in winter.

Many of my teachers were well into their 60s. Mr Testa had been in the Merchant Navy and told us that he no longer could smoke his pipe as his 'pipe tooth' had been extracted some years before. Mrs Ahlov played the piano. She had an outsized backside that wiggled as she strode across the classroom. However the chief sales lady in the Lilley and Skinner shoe shop in Golders Green had an even bigger bottom that seemed to fill the entire shop with its vastness.

Human society the world over has much to say about buttocks. There are at least 46 synonyms in the English language to describe male and female bottoms. Many are acceptable terms to describe this anatomical topography. Others are used as insults, or superlative adjectives, particularly when a more choice word eludes the speaker.

What are the buttocks' function? I know, that's a teleological question. Here's an answer anyway. It allows a fatty cushion to make sitting more comfortable. The large amount of fat also enabled corporal punishment to be inflicted without jeopardising structures beneath the fat. Buttocks of course feature large in humour and art. Research in Western heterosexual men has strongly suggested that a 45-degree angle of protrusion of the buttock and lower spine in relation to the higher spinal levels in women is both sexually attractive to the men and is associated with better pregnancy outcomes. Wearing high heels can exaggerate this angle, or to put it more simply, make the bottom stick out.

I have very little to say about male buttocks. If they are contained within clothing, I can happily disregard them, unless a 'builder's bum cleavage' assaults my field of vision. The reason for this nauseating spectacle, is that the wearer's beltline has drifted south because of a large amount of abdominal fat – likely a beer belly. Women's backsides, to my mind, however, are more than just a dash mysterious and unfathomable. They can be both highly erotic and yet at the same time disappointing, or even associated with some disgust. Seeing my mother's backside as she dressed was an embarrassment to me, giving rise to a moment of anger, thinking to myself, 'why can't she just tell me to wait outside for a few moments rather than plunge into undress?'. Maybe she was in a rush that day or had worries on her mind. My brain automatically flashes from that upsetting occasion to pictures of women, ordered to be stripped naked by Nazis, arms making futile attempts to cover up the shame of nudity just moments before they were shot and pushed into mass, anonymous graves.

Women's bottoms are at their most erotic when clothed in a well-fitting miniskirt, the narrowed waist expanding into the rump roundness, then tapering to shapely thighs. The late Victorian fashion of bustles gave the appearance of an enormous bottom, even where the real rump was of tiny proportions. Back to the mini-skirted woman. The outer or lateral aspect of naked buttocks contains a slight concavity or depression, which is masked by the skirt. Its presence without clothes is mildly disappointing. The 'bum slip', that is the fleshy underside curvature of the buttock cheeks are, in the nude woman, generally perceived to be highly erotic. They are more generally displayed nowadays by women wearing very short shorts and of course thongs. Women past their first flush of youth wearing small bikini bottoms on beaches reveal this bum slip area along with multiple creases – and why not – they and I should be proud of our bodies as we age! Some of us, however, wish to defy the aging process. The internet tells me that Gwyneth Paltrow exercises for 12 hours a day to keep her posterior toned. I am sure no one in ancient Greece had such tenacity. Nevertheless, I have never seen wrinkled bum slip areas portrayed in any of the nude sculptures I have seen in Greece or in the Victoria and Albert museum. That's male sculptors for you! Or maybe the problem lay with sexist male art patrons. And of course when in love these peccadillos become irrelevant.

Lewis, DMG, Russell, EM, Al-Shawaf, L, Ta, V, Senveli, Z, Ickes, W and Buss DM. Why Women Wear High Heels: Evolution, Lumbar Curvature, and Attractiveness. Frontiers in Psychology 2017 Nov 13;8:1875.

Back to the Lilley and Skinner shoe vendor. She spoke with a strong continental accent. 'Ve vill now Xray your feet to see iv zee shoes fit.'

X-ray machine for assessing shoe fit.

X-ray machines or to give them their proper title 'Shoe Fitting Fluoroscopes' were introduced in the USA in the 1920s and found their way into UK shoe shops some years later. The feet with shoes on were inserted into a space atop one or two steps, and one or more viewers could peer at the eerie, green illuminated outline of the feet within the shoes from one or more viewing portals at the top of the machine. A typical viewing in the USA would deliver 0.13 mSv of radiation over 20 seconds. If multiple shoes were tried on the dose would of course be cumulative. A modern CT scan of the head delivers 0.2mSv. From the 1960s onwards there was a growing awareness of the radiation hazards of these viewing boxes, which were finally withdrawn in the 1970s.

My primary school teachers were all both pleasant and supportive. I have thought over the years that both Mrs Ahlov and Mr Testa must have been young adults at the start of World War 1. An opportunity missed, to ask them about their war experiences. If I had had the presence of mind and courage to bring up the topic, I might have been rewarded with tales of bravery and sacrifice. The school was nominally Christian in outlook. The 1944 Education Act in England had stipulated that the school day for any educational establishment should begin with an act of collective worship. All pupils were also to be given regular religious instruction. Interestingly, the Act did not mention Christianity. It was left to the local authorities to come up with a syllabus. Most of them set out that Christian teachings, studying the life of Jesus as well as the Old Testament should be undertaken at assembly as well as in class. I became confused. Judaism at home and the synagogue and Christianity in school. When I confronted my parents with this conflict they told me just to disregard religious teaching at school. While browsing through a bookshop I chanced upon a pictorial life of Jesus. I asked my mother if we could buy it. She went ballistic, asking how on earth I could want such a book. More confusion! I guiltily enjoyed singing hymns at daily assembly, and learned the Lord's Prayer by heart merely by hearing it recited daily. The gathered assembly read the words of the hymns from two large books of sheets that were on pulleys and elevated for all to view. I still remember many of the tunes and words of these hymns as well as those of Christmas carols. A particular favourite among the children was 'Onward Christian Soldiers'. If I had known that tens of thousands of Jews had been massacred by Christian soldiers in the Crusades I might not have sung so heartily. I remained muddled and conflicted throughout my junior school days. As I saw it then and see it all the more clearly now, both my parents had narrowly escaped Nazi Europe with their lives because they were Jewish – and yet here I was, a misplaced Jew in a Christian environment. I guess they must have had their reasons. My brother also went to a non-Jewish primary school – not the same one as I.

At Christmastime I took part in two plays. One was a nativity and in the other I regaled my thespian abilities acting a frog! Strangely, I did not stammer in either part. Possibly this was because I did not have to think and speak at the same time, having learnt my lines by heart.

As I grew older I walked to and from school unaccompanied. There were three possible routes. Each was within the territory of one of three free-roaming dogs, which could appear unexpectedly to bark and snap at my short-trousered bare legs. Negotiating them was an exercise is fear (mis) managed. Nowadays it is illegal in the UK to allow dogs to be out of control in public places.

The Golders Green Estate was large enough to possess its own 'parade' of shops in Pennine Drive. The grocer-cum-delicatessen store was owned by a 'Mr K'. His slicked-back, shiny, jet black hair, pencilled moustache and sleek golden cigarette holder gave him the appearance of what was then known as a 'spiv', a term used in the 1940s and 1950s denoting a man who was a flashy dresser living by disreputable dealings. In his case his outward appearance was true to form. My mother had the measure of him, buying just what she needed and no more. I was, however, more gullible.

On one occasion she instructed me to buy a jar of instant coffee. 'A small jar only David, do you understand?' she stated. So off I went and meekly asked Mr K for a small jar.

'Your mother never buys small jars of coffee,' he retorted. I was confused. So I brought home an extra large jar – and got in big trouble with my mother. Wartime food rationing, and meat in particular, could be in short supply until rationing ended in 1954. Mr K however was sometimes able to procure special culinary items 'under the counter' for an appropriately exorbitant cost, but we generally could not afford his high priced delicacies. My mother had to therefore improvise. One of her 'meat' lunchtime dishes was chopped lung. At the time I did not know what I was eating. In general we do not care to think of the provenance of the food we eat, in particular animal products. I would certainly not eat chopped lung by choice. It is considered dangerous in the USA as it can be contaminated by the animal's other bodily fluid. Hence real Scottish haggis cannot be imported into the USA.

The shop in the parade that interested me most was the chemist, owned by a Mr Calcutt, a quiet spoken, balding man who peered over his half-moon reading glasses at me. His shop window contained large flasks of water, coloured red, yellow, green and blue. In later years he would supply me with the chemicals I needed to enhance the chemistry set reactions I undertook in my parents' back garden. These included magnesium strips that shone with a bright white light when ignited. Truly a 'chemist' shop rather than just a pharmacy! Chemistry sets in the 1950s contained substances that today are considered dangerous. One such reaction was to add liquid glycerol to a mixture of potassium permanganate and glucose. The mixture ignited spontaneously, producing an intense lilac flame confirming the presence of potassium. In the early 1950s in the USA 'Atomic energy' chemistry sets were available that contained uranium 238! A boy's out of school hours could be fun then but dangerous!

At school three of my fellow school pupils were particularly memorable. Celia Young was both clever and good looking with a dark complexion. I believe many of the boys in our class admired her for her beauty and mind, but silently from afar. Hilary Hodges had long wavy brown hair and was a graceful athlete. Terry Howard was good looking but did not seem interested in academic studies – however he was fabulous at all sports. I once went for tea in his house on the local council estate. Tea meant tea and bread and butter. Terry always beat me at playing conkers.

The game of conkers, using the seeds of the horse chestnut tree, is first mentioned in literature in Robert Southey's memoirs in 1821. The etymology of the word is not clear. It may derive from the word to conquer or be onomatopoeic by virtue of the sound one conker makes when hitting another. A conker is prepared for tournament by drilling a hole in its centre and inserting a string or shoelace with a knot to enable it to dangle. Some conker owners baked their chestnuts, which made them harder but more brittle. The competitors take it in turn to hit their adversary's conker with theirs. If it smashes the opponent's conker it becomes a 'oner'. If it goes on to smash another, a 'twoer' and so on.

Oh please you great horse chestnut tree
Tell the world when you saw me
When I was young and very small
And you were not so very tall

You wake me up, go back, remember
To autumns past, to mid September,
With mummy, daddy – Sunday ride
Come back with conkers and such pride

Hard brown shiny spheres array
In my hands and lap you lay
Waiting to be bored and stringed
Ballistic missile, conkering

Flying, fission, fracture, flee
I've broken Terry's – Victory!
Oh fantasy your lie is wide
A 'oner' yes but then you died

Homeward groans the car to tea
Why don't I have dexterity
To bash his boasty, browny ball
Hit Terry's 'tenner', and so he'll fall

Better that I do not enter
Competition and surrender
My conkers will not broken be
I'll keep them safe and sound you see

I'll store them in my cupboard drawer
Not through them a hole to bore

Or string them like washing on a line
But look at them from time to time

I'll caress them in December
And they'll remind me of September
Of warmer lighter halcyon days
Of autumn mists and country ways

Of orange red bespeckled clouds
At sunset and the drooping flowers
Of aeroplanes that fly and stray
Tween peppered embers of the day

Of hands held tight and loving arms
Of buttered toast and mummy's charms
Of warm and tender friendly nights
Oh conker ever gleaming bright

In spite of being thin to the point of skinniness I was keen to take part in sport at junior school. My asthma during those years was principally exercise induced, and could be severe. Running any distance over 100 yards resulted in marked shortness of breath and wheezing. If that happened at school there was little I could do apart from waiting up to two hours, in marked discomfort, for the whole episode to subside. If I was at home, I used a Rybar inhaler which contained ephedrine.

This worked by increasing the body's concentration of noradrenaline and so widened the constricted bronchial tubes. It certainly was not portable in the sense that one could take it to school, but gave relief in a few minutes at home. My GP, Dr Fishman was not keen that I used it, I think because it had the potential to cause cardiac rhythm issues and could also be addictive. When he came on home visits (GPs did that then even for 'flu!) it had to be well hidden. By the 1960s portable isoprenaline inhalers had been introduced, although they too were not really safe in terms of heart issues. When I was ten I suffered a succession of bad asthma attacks with bronchitis, likely made worse by the high levels of air pollution in the London at that time, in spite of the advent of the Clean Air Act in 1956.

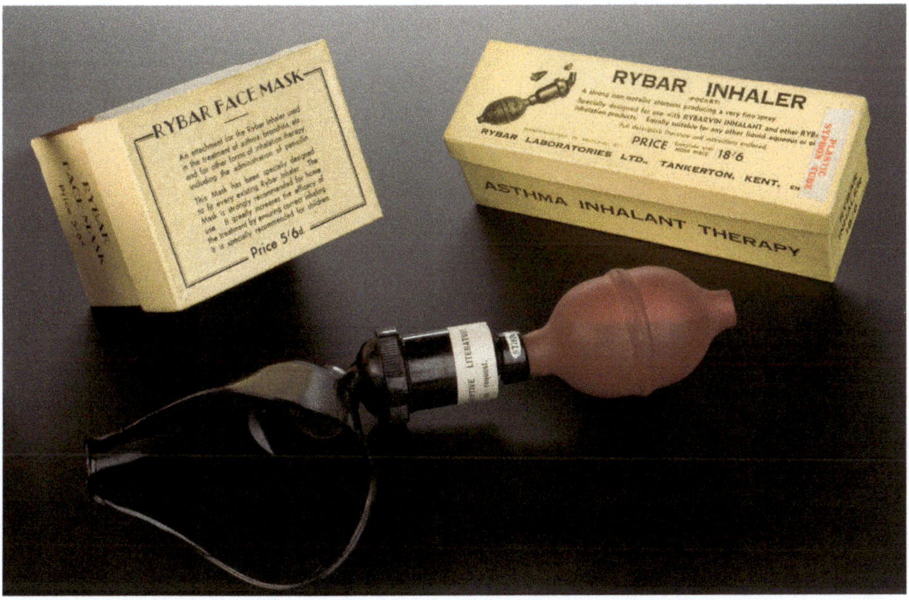

A Rybar inhaler (Courtesy of Science Museum Collection)

The Clean Air Act prohibited 'dark smoke' from being emitted from either home fireplaces or industrial furnaces. The Act also prohibited grit and dust from being emitted. Chimneys had to be high enough so as not to prejudice the health of local residents. The darker the smoke, the more polluting it was deemed to be. The darkness was measured using a Ringelmann chart for assessment of the colour of the smoke. In practice, this meant burning coke instead of coal. Coke is produced by heating coal in an airless kiln up to 2,000 °C. The process vaporises or decomposes organic substances in the coal, driving off coal gas and coal tar. The porous coke has a high carbon content, so allowing it to produce heat with a minimised smoke production.

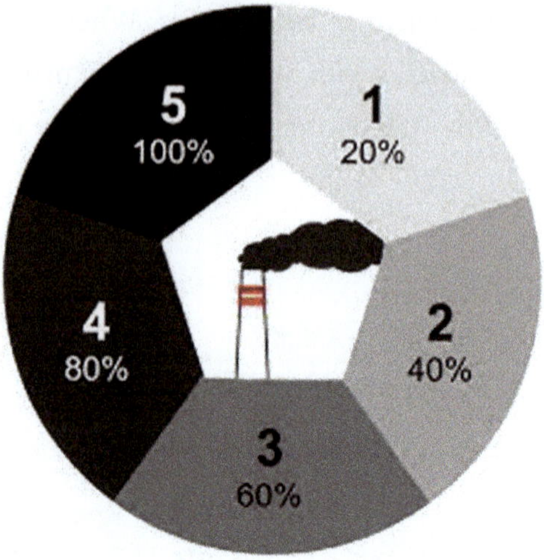

A Ringelmann chart[1]

A particularly severe and prolonged asthma attack was spontaneously brought to an end in the clean air of Switzerland, where the family took the first of our many summer holidays. A neighbour taxi driver ferried us, at 1 am, to 'London Airport' as Heathrow was known in the late 1950s. My excitement at the anticipation of journeying to a foreign land was richly rewarded by seeing an orange red sunrise over the Swiss Alps as the large, turbo-propeller driven, oval windowed Vickers Viscount descended humming to Zurich airport, from where we transferred to the Hauptbahnhof. Continental train platforms are at ground level unlike those in the UK which are at train level. A powerful (5770 horse power) Ae 6/6 red electric loco powered our train over the Alps, rising by means of multiple scenic loops and short tunnels to a height of 5,000 feet, finally plunging into the old, 9 mile St Gotthard tunnel (there is a nowadays a new, lower level St Gotthard tunnel – efficient, but less scenic, used by both cars and trains to cross the Alps) at the end of which it burst into to the bright sun-kissed canton of Ticino and our

1 Diagram appears in Prakasa, E., & Muchlis. (2017). Development of imaging based method for plume opacity measurement. In 2017 5th International Conference on Instrumentation, Control, and Automation (ICA), pp. 212–216 and Hughes, G., 2010. The Ringelmann Smoke Chart, International Initiative for Impact Evaluation.United States of America. Retrieved from https://coilink.org/20.500.12592/dc9fcs on 13 Sep 2024. COI: 20.500.12592/dc9fcs.

final stop at Lugano, a large Italianate town nestling by the tranquil Linden tree lined Lake Lugano, surrounded by pine tree covered alpine foothills. The clean mountain and lakeside air soothed my bronchial tree, only for it to be re-irritated on our return to grimy, suburban, London.

Dr Fishman and his family lived at 26 Pennine Drive. We lived at 22. His was a single-handed practice, which meant being available 24 hours a day, 7 days a week, in the main undertaken from his consulting room attached to the house with a narrow waiting room adjoining. Rarely was anyone else waiting, so you just stared at the empty fish tank or read the out of date magazines. Right up until he retired in the early 1970s, there was no appointment needed – patients just pitched up, waited their turn if necessary and then walked in to see him. He sat behind an enormous wooden desk piled high with papers. To his right was an examination couch, X-ray viewing box and on the left a white metal and glass instrument cupboard and small steam steriliser. The days of disposable, single-use needles and syringes were yet to come. If an injection was needed he would pick a glass syringe and needle from the steriliser with large forceps. The needles' ends inevitably were blunt, with the tips microscopically but definitely bent by previous use. This meant that their passage through skin and muscle involved forcing a hook through the tissues rather than a less painful clean cutting of a once only used sharp needle. Dr Fishman was brought up in Tiger Bay Cardiff and went to medical school at Barts in the 1920s. Fast forward 15 years, to when I was at medical school, where I heard the pejorative but humorous phrase 'You can always tell a Barts man, but you can't tell him very much!' When I knew Dr Fishman, he had thinning greying tight curly hair, a moustache and a slight stammer. He pipe-smoked a very pungent aromatic tobacco, so that the waiting room and his consulting room were awash with the mixed odour of his pipe tobacco and carbolic soap. The X-ray box had the same X-ray permanently showing, which in fact was of my chest! Seen from the 21st century the surgery was a medical museum, with its antiquated equipment which included an ultraviolet light for the treatment of vitamin D deficiency. I recall having to strip to the waist in his icy surgery to have this light shone on me while I shook and shivered. Many years later I did one locum session for him. I sat rather proudly in his chair for two hours but saw no patients in that time!

In 1955 the first, highly effective injectable polio vaccine, made by Jonas Salk, became available in the USA. It was unobtainable in the UK till the following year. My Auntie Betty – my mother's sister- was a nurse who ran a care home in upstate New York and was able to obtain a supply for my brother and myself. Dr Fishman agreed he would give the injections but only on condition that two more sets of vaccine would be sent – for his two children!

Polio, or to use its full name, poliomyelitis is a viral disease, caused by a gut or enterovirus, which passes from the bowel wall into lymph glands, and thence to the blood, ending up in spinal cord motor cells i.e. the ones which transmit messages to muscle, resulting in flaccid (low tone) muscle paralysis. It is transmitted via the faecal-oral route. Although only 1 per cent or less of those infected develop muscle weakness, it killed over 50,000 people a year in the USA and proportionately as many in Europe prior to the vaccine, often because the respiratory muscles were affected. President of the USA to-be Franklin Roosevelt, was permanently paralysed from the waist down by the illness aged 39. Polio was first described in 1789. In 1954 Jonas Salk, a virologist, prepared a vaccine by growing the virus in monkey kidney cells and then sterilising it with formalin. It was administered by injection. Both the observational and the placebo controlled studies undertaken in the early 1950s in 5–9 year olds demonstrated great success at preventing severe illness and death. In 1970 only ten people died in the USA of polio. The Salk vaccine was superseded by the oral Sabin vaccine which was easier to administer and just as effective.

Monto, AS. Francis field trial of inactivated poliomyelitis vaccine: background and lessons for today. Epidemiologic Reviews 1999; 21(1):7–23.

Back to school sports – I was also quite proficient at high and long jump and won the sack race at a school sports day when I was nine. However, I was a poor soccer player, continually being hectored and shouted at by players in my team for not passing accurately or not hitting the ball into an open goalmouth.

Bullying was not uncommon at school. I was punched in the face by a group of older boys after being surrounded, harangued and asked if I was Jewish. Children, of course, are not naturally anti-Semitic but may become prejudiced after hearing racist adult remarks. There had been widespread anti-Semitism in the UK in 1947, particularly after the Irgun in Palestine killed two British soldiers in reprisal for the hanging of Jewish militants. Anti-Semitic riots in the UK continued well into the late 1950s, as well as a resurgence of more organised anti-Semitism by fascist groups. Being the child of heavily accented German refugees parents I guess added to my tormentors' xenophobia. The loss of 6 million Jewish lives in the Holocaust was not an open topic for discussion even amongst Jews in the 1950s, and had certainly not percolated to any extent into the average UK household, who were understandably much more concerned at the time with the threat of nuclear annihilation, *Rock Around The Clock* sung by Bill Haley & His Comets and the more immediate and obvious racial disharmony in the Notting Hill riots involving violence against the local the Afro-Caribbean population. The fascist Oswald Mosley, who had been imprisoned during the war because of his pro-Nazi and fascist rallies, was able to re-enter politics briefly in the 1950, retaining his extreme right-wing views which found sympathetic ears in some quarters. It distresses me that anti-Jewish feeling continues to raise its head in the UK in the 21st century. *Plus ça change, plus c'est la même chose!*

Clitterhouse school (the name Clitterhouse probably derives from the 13th-century word clite meaning clay, which was used, presumably, to build the original farmhouse after which the adjacent school was named) had a three form entry – about 35 in each class. I came to understand that there was almost total inability to read in the lower two streams, even by age 11. I am unsure how to explain this – mass dyslexia or ADHD – neither seem probable. I feel it may have been a combination of poor teaching along with an ethos of homes not having books available, but rather picking up information from the radio and television, although the latter was only available a few hours a day in the 1950s.

The school buildings were adjacent to a large field that sloped gently down to Victorian terraced housing, built for families of men who worked in the London Midland railway that lay just beyond their red-bricked homes. Playing in the field was a carefree lunchtime adventure, running in the long, dandelioned, bluebelled grass to the auditory accompaniment of engines hissing and clanking into goods wagons in the nearby railway marshalling yards, and the unforgettable sulphurous odour of steam and smoke. Fifty years later I visited the school on the off-chance of there being any records of my time there. The current headmaster found a logbook for 1958 written by my headmaster Mr Pearson, describing a school trip for the top year to visit Roman St Albans. I have a hazy memory of that school outing. We trooped off to Cricklewood Midland station, a 10-minute walk from the school, and boarded the slow, local steam train up line to St Albans. Mr Testa warned us not to lean out of the carriage window, citing a cautionary tale of a man who had done so resulting in him being decapitated as the train entered Elstree tunnel! My only other recollection of that outing is that a number of my classmates wrenched pieces of masonry from a Roman wall to take home as keepsakes!

My psychological health while at junior school remained problematic. I was fidgety and nervous in class but generally more at ease at home except when my mother and father went out together to visit friends on weekend evenings. I then became highly anxious and tearful. Running through streets seemed to help. I had so much of my own adrenaline on board that I did not develop asthma. I was convinced my mother would never return. Strangely if she went out on her own there was no problem. These episodes must have been panic attacks, and may well have been related to my earlier experiences of abandonment in hospital. Other possible pertinent issues were my father's rare but fierce temper. He would then shake me violently and smack me round the head. I have a memory of occasions he only just held himself back from hitting my mother. She in turn threatened to leave him a number of times but didn't because of her children. I do not think they were well suited as a couple but stayed together over the years to preserve a semblance of family unity. I guess I might have felt I could not trust him when they went out together. These fearful episodes reached a shameful climax when they went to a West End theatre with my aunt who had come over from the USA. In desperation I phoned the theatre asking for my mother to be tannoyed. She came to the phone a few minutes later and understandably was very angry and embarrassed that her name had been publicly announced. I too was very ashamed and upset that I had disturbed their evening. I guess the lengths that I went to in order to determine her safety were a measure of my high level of distress. The panic attacks were a problem from the ages of 8 to 13. Having a 'babysitter' only marginally helped me to cope. Mrs Bertel was one such person who lived near us. She was a small, quiet spoken woman in her mid 80s. I listened to her tales with some interest and distraction. She was very insistent that she had taught her sons to swim as children. One of them had been in the British Expeditionary Force at the Dunkirk troop evacuation and would not have been able to reach a Navy destroyer and England, had he not been able to swim out to the small boat that picked him up from shallower waters. To return to the misery of my panic attacks – I was sent to a child guidance clinic in Paddington where a psychodynamic approach was undertaken – but it didn't help. I seemed to grow out of the problem in my early teens. All sounds very oedipal! My father's behaviour towards me as detailed above paints a picture of him as an ogre. In fact most of the time he was pleasant and cared deeply for his family and others. I am pretty sure he saw domestic violence in his own parental home. His father lived on till his early 90s living in Stamford Hill. My father was pretty useless at mending even the simplest broken machinery and would soothe himself by making Turkish coffee.

There was really no doubt that the television set was non functional. But my father would not accept its demise. His attempted resuscitation by repeatedly manipulating the on-off switch and plugging the electric source to another power point. He was about to miss the weekly current affairs programme. But disappointment turned to hope as his groaning filled the room and then evaporated into a sigh.

'Let's make some Turkish coffee, and I'll phone Mr Tuwie'.

He rummaged around the kitchen until he found the small brass boiling cezve. He passed it to me while he lit the gas. It felt smooth and cold. I filled it with water and place it on the range. Finely ground dark brown coffee was loaded by the tablespoonful into the boiling water accompanied by liberal helpings of sticky demerara sugar. We didn't have long to wait for the heated mixture to suddenly bubble and froth. With each eruption came the mixed aroma of wood, fermenting fruit and tobacco. My father reboiled the mixture three times and then decanted the supernatant coffee into the blue and gold decorated miniature cups. The coffee was rich, thick and almost sickly sweet and seemed to satisfy and pacify my father.

Mr Tuwie came to look at the television the next day, even though it was a weekend. He was a tall, swarthy man whose long face was topped with a halo of erect coarse black curly hair. He not so much spoke but rather bellowed, 'WHAT'S UP WITH IT?'. I was both fascinated and somewhat disgusted by both his dense aromatic body odour, as well as the small frothing saliva balls that danced at the corners of his mouth as he spoke. He unscrewed the back of the television, commenting on his findings as he went along. 'VALVE GONE HERE. IT FEELS ROUGH. THEY DON'T MAKE THEM LIKE THEY USED TO'. He replaced it, and switched on the power. Hissing white noise filled the room. He replaced the coverings. 'TRY IT'. My father's face shone as the TV came to life.

My father's pickle business flourished during the 1950s. It was centred in Hardinge Street, between the Commercial Road and Cable Street in London's East End, which at that time consisted of slum dwellings amid swathes of brick and concrete rubble where Nazi bombs had flattened the area. Pinkish red wild flowers grew on those seemingly barren sites. I later found out that these were Rosebay Willowherb, a very hardy and often a 'pioneer plant'. The ugly 1960s high rise blocks had yet to be built. Hardinge Street, in common with much of the surrounding area, has become gentrified in the last 20 years, reflected in soaring property prices. The proprietors of the pickle business were my father and his younger brother Bruno. Bruno was an excellent businessman and did most of the buying and importing. By any standards he was a strange man, working on the phone through the evening and into the night. Dawn was his cue to put work on one side. Initiating sleep eluded him and so he found a compliant doctor who prescribed the dangerous barbiturate hypnotics. By mid afternoon he was awake, and the cycle repeated. He continued to smoke cigarettes through the 1950s and 1960s, by which time my father had stopped. Bruno had a marked anxiety related tremor which exacerbated his very unattractive and messy eating style. He had remained on the family farm until uprooted and fleeing to England in 1939. Unlike my father, who had cultivated a certain civility and panache as he travelled round Europe, Bruno seemed to have retained the unsophisticated demeanour and mannerisms of a German peasant farmer. That said, he was good hearted and tried his best to play with my brother and myself. He remained single all his life and lived on his own in various flats in and around Swiss Cottage Hampstead, among a thriving Austrian and German Jewish refugee community. When not at work or asleep, he frequented the Cosmo restaurant in the Finchley Road, which from 1937 onwards was a meeting point for German speaking refugees. Many were intellectuals, including Sigmund Freud, who lived just round the corner. In its heyday the Cosmo was alive with hot debate, smoke and the aroma of fine coffee. It closed in 1998. Bruno died in 1964 of an accidental overdose of the barbiturates mixed with alcohol. He was the first dead person I ever saw. His flat was icy cold. It was as empty of life as the corpse in front of me. No furniture to speak of, no food in the kitchen. Just an empty bottle of brandy and a bottle of pills by his bedside. After his death my father sunk into a prolonged deep depression, the business eventually folding.

The Cosmo Restaurant

But back to happier times in the 1950s. My father worked on the sales side of the business. Chefs of both prestigious hotels and restaurants as well as less salubrious establishments could pick from any of a number of wholesalers who sold almost identical culinary products. Many wanted to be given personal financial inducements to ensure the passage of a sale. My father often complied with their wishes. The factory was housed in three adjacent terraced houses. Their sloping walls and rickety wooden staircases would nowadays be considered unsafe for human habitation, to say nothing of the food hygiene. The acrid odour of pickles had soaked into the very fabric of the dwellings – even the offices. Food production took place in a large, uneven, concrete floored backroom, illuminated by an overhead glass skylight plus a few naked light bulbs. Ten local women did the packing and pickling. The cacophony of their shrieking voices and laughter echoed around the smoke-filled room. With multicoloured knotted headscarves and cigarettes hanging from their lips they greeted me ''ere's the governor's boy, don't be shy young'un'. In later years mayonnaise was manufactured in house with up to date industrial equipment. Each weekday my father drove himself and a friend, who worked in the furniture business in Hanbury Street, into the East End, leaving home at 6 am. I spent the occasional day there talking to the secretaries, helping deliver goods and going to the London docks to pick up produce from abroad. The docks in the 1950s were situated in the main in the East End of London and covered a large area north and south of the Thames. They were heavily bombed during World War 2 and closed in the later part of the 20th century, being replaced by large, deep water docks such as Felixstowe and Tilbury.

Weekdays my father left before we children got up in the morning and arrived home well after our supper.

He encouraged us to ride two-wheeler bikes and bought them for my brother and myself not long after my eighth birthday. My brother and I were friends with the GP's son David Fishman. His parents would not buy a him a bike believing it was a dangerous and unnecessary piece of machinery.

A paper in the British Medical Journal in 2020, plus an accompanying editorial discussed the pros and cons of commuting on roads by bike compared to other less active modes of travel. 230,000 commutes were analysed. Compared to non-active commutes such as bus or train there was a definite increased chance of injuries to arms, legs, head and trunk in the cycle commute group. However the cycle group had a lower level of cancer, cardiovascular disease and death compared to the non-active participants. The editorial advises the best of both worlds is to have dedicated cycle lanes.

Lusk, A. Designing better cycling infrastructure. British Medical Journal, 2020; 368 :m848 doi:10.1136/bmj.m848.

We therefore had access to two bikes between the three of us. So began the search for a third mode of transport. In the 1950s people disposed of their unwanted household goods rather more willy-nilly than nowadays, although 'rag and bone' men, with a horse and cart took away much unwanted household material which they either mended or sold on to metal merchants. We three therefore scanned a local 'dump,' where odds and ends had been fly-tipped, and found the chassis of an old pram with handle intact. Placing a makeshift wooden seat on top completed the carriage. We attached the 'cart' behind the saddle of one of the bikes. One of us pulled the cart and passenger – this needed quite some effort by the cyclist, particularly when David Fishman was the passenger, as he was quite heavily built.

When I was eight I was taken to an air display at Hendon aerodrome. Multicoloured aerobatics, low level planes of yesteryear as well as up to date jet fighters screaming just metres overhead thrilled a small boy! Hendon aerodrome was in use as a commercial and RAF airfield from 1908 to 1968. The UK's first airmail flight, first parachute jump from a plane and first night flight all took place from there. During World War 2 it was mainly used for transport activities. After 1968 most of the land became the Graham Park housing estate. The former site is also occupied by the RAF Museum and the Hendon Police Training College.

That same year (1956) the family went on summer holiday to Knokke, a Belgian seaside resort close to the Dutch border. The topography of the resort is no higher than 10 metres above sea level and very flat. It was one of my early tastings of continental Europe. We flew in a DC3 plane from Lydd in Kent.

The DC3 or Dakota was first produced in the USA in the 1930s and became the workhorse transporter in World War 2. 16,000 were built. When used as a passenger plane it could fly about 30 people and was powered by 2 piston engines. The cabin was not pressurised. Our flight was short, just crossing the Channel at low altitude. However it was a rough ride that day and I felt terribly sick and developed bad earache on the descent. We landed at Ostend, just 25 miles from Knokke. New sensations awaited me there, including smelling and tasting real coffee, as compared to powdered instant that I was used to, as well as great ice cream and tricycles that we sat on in recumbent position that seemed to us to speed along the flat beach promenades. I heard and reheard Doris Day's popular song of that time *Que sera, sera*, which I associate with the two weeks I spent in Belgium. Nearby was an exhibition of concentration camps photos. My parents debated whether my brother and I should view the exhibits in spite of our tender years. After some whispered discussion between them, they decided we should view the exhibits. What I saw greatly shocked me, in particular the bodies of emaciated children. Until that time I had not understood the horror of the Holocaust.

Yad Vashem, meaning a name and a memorial, is Israel's official site for the remembrance of the 6 million Jews who perished in the Holocaust. It lies within the greenery off Mount Herzl in the Jerusalem forest. There is a dedicated monument to the 1.5 million children killed by the Nazi regime. The main room of the underground children's memorial is enveloped in darkness. In the centre is a single burning candle, the light of which is reflected by many mirrors, transforming the one candle flame into hundreds of points of light. This phenomenon has been described as a 'congregation of tiny souls in search of bodies'. It is easy to imagine that one is in the company of those 1.5 million children. Background noises are stifled to a spine chilling silence. The only sounds in the light speckled room are the hushed spoken names and places of slaughter of all the martyred children, recited on a looped tape, one by one. No child's name is missed over the three months it takes to voice all the 1.5 million names. Then the looped tape begins its memorial run again.

Aged 11 all of my school year took the so-called 11+ examination to determine whether we would be sent to a grammar or secondary modern school – the latter having a lower level of academic input. My brother Michael was already at Orange Hill Grammar School in Burnt Oak near Edgware.

When I felt energetic my performance in English, maths and general intelligence tests was good. However I often felt tired and think that I only achieved borderline status on the day. In spite of this I achieved my goal of getting into a grammar school.

Chapter 4

TEENAGE YEARS AND SCHOOL

So in September 1959 off I went to Orange Hill. The school motto was 'Omnia probate, bonum tenete' (try out everything but stick with what is good).

In that spirit I ventured onto the 226 bus which stopped in Pennine Drive and ended at Golders Green Underground Station.

RT1, AEC Regent III prototype, in the London Bus Museum
Credit: Murgatroy49 Licenced under Creative Commons Attribution, CC BY-SA 4.0, https://commons.wikimedia.org/w/index.php?curid=131362157

I had a prepaid season ticket each term. The 226 was a red, double decker AEC (Associated Equipment Company) Regent model which ran many routes in London at that time. The 9.6 litre diesel engine pounded out a baritone rhythm, shaking the bus with each cycle. The uniformed conductor was usually a middle-aged woman with short black curly hair, who ensured everyone got on and off the bus safely and in orderly fashion. She announced, 'Hold very tight please' and 'Fares please' in BBC style 'received' English. I wondered if she had started to work for London Transport during the war years, when women did many jobs previously seen as all male enclaves. She carried a narrow wad of different coloured tickets, from which she would pull one and clip it. She kept the cash in a large leather wallet strapped round her shoulders.

The late autumn of that year, 1959, a thick 'pea soup' London smog engulfed the 226 on its way to Golders Green. Smog consists of smoke plus fog. The bus halted as we entered the brown yellow haze. The driver came out of his cab and spoke to the conductor. She agreed to walk a few yards ahead of him as pathfinder, swinging her torch from side to side. We perceived her effort as a faint yellow light, aimlessly roaming amid the smog. At this walking pace we eventually reached Golders Green an hour later, only to find that all the Underground trains had been cancelled.

The AEC bus possessed no passenger doors, rather an opening at the back left with an upright pole to cling on to. There was no heating to speak of in winter and it was stuffy in hot weather. The agile and brave would get off the bus while it was moving but importantly facing in the direction of travel to ensure a safe landing.

At Golders Green I transferred from bus to the above ground section of the Northern Line Underground to Burnt Oak. In 1959 the carriages were 1939 stock, which were the first underground trains to have their motors fully under the carriages, so making more room for passengers. The floors were wooden and slatted and sloping towards the doors on both sides, so enabling the moisture not to accumulate on the surface. The trains did not have the power of modern deep Underground stock. Going uphill from Brent Cross to Hendon Central, the train's motors groaned in displeasure. Colindale was the last but one stop for me. The Police Training school and car skid pad were visible from the train just before Colindale station. I got off at Burnt Oak and walked 10 minutes up Orange Hill Road.

The school was built in the 1930s on Abbots Road which is on a steep incline. The boys and girls sections were adjacent from each other but completely separated. I made a good friend for life in the early years there – David Lewis who had a burning desire to study medicine, some of which rubbed off onto me. Television programmes of the day, such as *Dr Finlay's Casebook* set in the Highlands of Scotland and *Dr Kildare* set in the USA, also whetted my appetite. David Lewis has a photographic memory which at the time I did not quite understand, in that homework that took me hours, took him a fraction of the time. In spite of his academic ability he was always very generous helping others with their homework and humble to boot.

Many of the teachers at Orange Hill had their personal foibles and idiosyncrasies. Others were bullies, inflicting corporal punishment – the cane – and seemed to me to be lords of all they beheld in the classroom.

Mr Willett taught French. He was a short, stocky man in his 40s with wavy black hair, who caught my attention for a number of reasons including his almost continuous bottom scratching, and for singing Christmas carols in French. A subsequent French teacher was Mr Reffold, a mean-looking, bespectacled Oxford graduate with a psychopathic personality disorder. He ordered silence in class on a day that I guess he felt he had an urge to act out his callousness and brutality. I had dropped my pen and asked someone to pick it up as it fell away from me. For this minor peccadillo I was heavily slapped round the face, the result of which was he broke my eardrum. I think I cried out, 'I can't hear', at which point he dragged me to the headmaster's office. The headteacher took a dim view of Reffold's actions. Nevertheless, Reffold continued teaching. Nowadays he would be charged with a criminal offence. He later went on to another school where he was stabbed in the chest by a pupil, later leaving teaching altogether. I traced an obituary for him to his old school magazine. His fellow pupils strangely also referred to him as 'Reffold'. He died in middle age. Miss Buckle the art teacher, thought my paintings 'romantic'. Mr Rivington taught English. He was a tall, slender, bespectacled man, who, we were told, had come from the Rivington publishing house and made no secret of his criticism of the school end of term hymn *Jerusalem* (which I always enjoyed listening to) saying that the composer of the words, William Blake, would have been very unhappy with the tune by Hubert Parry. He also introduced us to the poem 'Macavity: The Mystery Cat' by

TS Eliot – a marvellous piece of tongue in cheek literature! Mr Harradine was a blond-haired, chain-smoking, newly graduated chemistry teacher who had been to Orange Hill school himself as a pupil some years before. As an introduction to chemistry he informed us about phlogiston, which was a 17th century notion that as substances burnt they released the so-called phlogiston leaving the dephlogisticated residue e.g. ash – a theory not extant for two centuries or more – so I wonder why he taught us this! Mr Rumble taught A-level physics. He was tall, corpulent, with a red bulbous nose and profoundly deaf. At the start of each week he issued us with a worksheet, which informed us which experiments we should undertake and which theoretical questions we should do as homework. Having done this he re-entered his sanctum and told us to get on with it!

A word about the headmaster Kenneth Howse. He was universally known as 'the dook', a London suburban diminutive, I believe, of 'the duke'. Tall and well built, with thinning straight hair, he would muse on what a pupil would tell him, as he gently chewed an arm of his spectacles. If a misdemeanour had been committed, he would announce to the whole school at assembly that he 'had names' and that punishment would be less severe if the culprit owned up. I think we all saw through that one! Nonetheless he was greatly respected by staff and pupils alike. As there was a shortage of space in the school, many pupils worked more or less silently in the corridors during free periods. When Mr Howes's squeaking shoes were heard approaching, the slovenly corridor learners would suddenly become ramrod straight and the prior quiet chatter devolved to a silence where one could hear a pin drop. No one knew exactly what Mr Howse did in World War 2. Some thought he had served in the Merchant Navy. Another version that he captained a minelaying ship in home coastal waters. If the cane was to be administered the offending pupil had to fetch it from Mr Howse's room, as well as the black book that came with it wherein was inscribed the offenders name, crime and the teacher who administered the punishment. The boy on whom the punishment was to be inflicted walked with the cane and book back to the classroom. As each classroom had windows on the corridor side the punishment was all the more humiliating as it was viewed by the gathered morbidly interested corridor learners.

There was strictly no mixing of the boys and girls school in or near the building. That said, I often saw a girl I thought was remarkably attractive at Burnt Oak Underground station. She always seemed to be by herself. I must have been all of 15. She had long light brown wavy hair and stood with her back slightly arched. I longed to see her each day and thought about her frequently. I never exchanged a single word with her – too shy – just admired her from afar! My ruminations and daydreams over her were generally of a romantic and rather than a sexual nature, holding her hand and maybe walking together somewhere other than Burnt Oak and Orange Hill Road.

CASE HISTORY – ROMANTIC LOVE – CIRCA 1990

Primrose was a 35-year-old academic mathematician. Whenever I saw her, her frizzy hair was tied back in a ponytail. When she first consulted me she had just acquired primary genital herpes, and was highly distressed by the pain and the psychological implications of the infection. Fortunately, over the next six months she did not suffer any obvious recurrences. However three months after I first saw her I happened to be in the walk-in clinic again. She spotted me and asked to see me again. This time she had moderately severe lower abdominal pain which turned out to be caused by gonococcal infection of her fallopian tubes. Her one and only ever sexual partner at that time was a man she had been seeing from time to time. He was employed at her university as an electrician, and was a good five years younger than her. She thought about him virtually all day long, and felt elated when fantasising about him, but knew that he was seeing other women. They had sex when it was convenient for him. However she was always ready to put whatever she was working on to one side to see him. I saw her for the last time two and a half years later when she attended the herpes clinic. By this time she told me she no longer was seeing him. She had 'fallen out of love' with him a year before and from her current perspective didn't really know what she saw in him in the first place. She had never been in love before. She was currently going on a few dates with a fellow mathematician but did not have such strong feelings about him.

As in the case of Primrose, romantic love can be seen as an aberrant mental state, which may seem to override any rational decision as to whether or not to have intercourse. It is almost as if freewill has been taken away from the person. Love expresses itself in many art forms. The Habanera (L'amour est un oiseau rebelle) *in Bizet's* Carmen *and Puccini's* Nessun Dorma *express love in the musical form. In the written form, Robert Burns's poem 'A Red, Red Rose'. In painting, romance may be expressed by tones of light and shade, symbolism and the depiction of a love story.*

A number of more scientific authors have discussed romantic love. The psychologist Dorothy Tennov, coined the term 'limerence' to describe an aspect of romantic love that is more transient and superficial than true love but nevertheless involves a profound infatuation, fantastic longing and seemingly deep obsession with another person. The biological anthropologist Helen Fisher seems to feel romantic love is more an 'infatuation'. Romantic love is not necessarily related to heightened sexual feelings, but the two are usually coexistent.

In the first intense stage of romantic love, there is an altered mental state, with intrusive thoughts of the loved one, an increase in energy, alterations in appetite, sleep and decreased rational judgement. There is also some evidence of increased bodily concentrations of the neurotransmitters noradrenaline, endogenous opiates and dopamine and a significant decrease in serotonin. This initial phase of romantic love thus probably shares some biological substrates with anxiety and fear. A young man has been described who developed a severe obsessive-compulsive disorder (OCD) (e.g. over checking and other non-partner related ruminatory activities) each time he fell in love. The implications of this is that low serotonin is involved in both states. Platelets in the blood also carry serotonin which is significantly lowered in both patients with OCD and in the early stages of romantic love. Other evidence that seemingly supports this notion of decreased serotonin being implicated in romantic states is that men on SSRI antidepressants – which raise serotonin in the body and brain – feel less romantic love and attachment once on the medication compared to before they started it. Romantic love seems also to be associated with raised cortisol and the brain peptide kisspeptin. These changes mostly revert to normal after the period of intense love abates.

Brain MRI scans of those in romantic love, when viewing pictures of or thinking about the loved one, have very distinctive features in both the cortex and subcortical areas that are quite similar to brain changes in people being successful at video games. Conversely brain MRI scans of people who have been in love but have been rejected show similar findings to people in the throes of cocaine withdrawal.

The attraction and infatuation aspects of romantic love usually last for up to two years, after which there is a gentler 'attachment' phase which is said to be more commonly associated with increased oxytocin – the bonding peptide. This is consistent with the modal time to couples divorcing being about four years – two years romantic love and two years romantic attachment. In terms of teleology, what is the point of romantic love? Evolutionary explanations might include initiation and maintenance of male–female relationships long enough to ensure reproduction and survival and early upbringing of offspring.

'My Love is Like a Red Red Rose' sung by Eddi Reader https://www.youtube.com/watch?v=jXH9DVS76yM

Marazziti, D, Palermo, S and Mucci, F. The Science of Love: State of the Art. Advances in Experimental Medicine and Biology. 2021;1331:249–254. doi: 10.1007/978-3-030-74046-7_16. PMID: 34453303.

Marazziti, D and Stahl, SM. Serotonin and Love: Supporting Evidence From a Patient Suffering From Obsessive-Compulsive Disorder. Journal of Clinical Psychopharmacology. 2018 Feb;38(1):99–101.

Bartels, A and Zeki, S. The neural basis of romantic love. NeuroReport. 2000 Nov 27;11(17):3829–34.

Marazziti, D, Akiskal, HS, Rossi, A and Cassano, GB. Alteration of the platelet serotonin transporter in romantic love. Psychological Medicine. 1999 May;29(3):741–5.

Marazziti, D, Akiskal, HS, Udo, M, Picchetti, M, Baroni, S, Massimetti, G, Albanese, F and Dell'Osso, L. Dimorphic changes of some features of loving relationships during long-term use of antidepressants in depressed outpatients. Journal of Affective Disorders. 2014 Sep;166:151–5.

Reyes, RP. Is love possible without Kiss1-es? Journal of Neuroendocrinology. 2015 Sep;27(9):739–40.

Fisher, HE, Brown, LL, Aron, A, Strong, G and Mashek, D. Reward, addiction, and emotion regulation systems associated with rejection in love. Journal of

Neurophysiology. 2010 Jul;104(1):51–60.
Fisher, HE. Evolution of human serial pairbonding. American Journal of Physical Anthropology. 1989 Mar;78(3):331–54.
Tennov, D. Love and Limerence. Maryland: Scarborough House, 1979.

Aged 13 my bar mitzvah was celebrated at Dunstan Road Synagogue in Golders Green.
The etymology of the word bar mitzvah, bar (Babylonian Aramaic for son of) and mitzvah (Hebrew for commandment) has the generally understood meaning of a boy of 13 reaching the age of religious majority, when he becomes responsible for his moral, ethical and religious actions. These include putting on tefillin (phylacteries) on a daily basis and being able to be included in a quorum of men praying. The Bible hints at this coming of age. However the occasion and the ceremony is only clearly described by the rabbis of the 14th century. The boy is taught to read a section from the scroll of the law using special cantillation or notes. The scroll contains the whole of the five books of Moses, but without vowels or music notation, in effect meaning it must be learnt by heart. There is usually a celebratory event after the reading.

Dunstan Road synagogue was erected in 1922, some 15 years after the Northern Line Underground had been extended to then green fields of Golders Green. The building continues to function as a synagogue to this day. It is a listed Grade II English Heritage structure.

My bar mitzvah reading from the scroll of law described the Amalekites harassing the Israelites in the desert. I also read/sang the accounts of Deborah the prophetess in the pre-monarchic Israel, who inspired a mighty victory against the Canaanites. I was taught this by the synagogue's cantor Reverend Tachlisky who had come to the UK as a refugee from continental Europe before World War 2. He was born in Ukraine and studied music at the Budapest High School of Music before moving on to cantorial posts in Berlin, Vienna and finally England. He had a rich and powerful tenor voice. My parents told me in whispers that his wife, who I recall as a tacit, rather nervous woman with long, black wavy hair, was a concentration camp survivor. This was not a topic for further discussion in those early post-war years.

The haunting photograph of a smiling eight-year-old Meir Lau taken in July 1945 in a post-war displacement camp conceals the traumas he suffered during the war years in concentration camps. He went on to become Chief Rabbi of Israel. He asks why Holocaust survivors remained silent about the horrors of the Holocaust for decades and concludes that 6 million was too large a number of deaths to comprehend. He discusses the reason why the Diary of Anne Frank was so popular during those early hushed years. The book does not discuss tortures or killings. He feels the reason for the book's success was that it was one person's story. One can comprehend one person's story. The memorial for the Holocaust in Israel, 'Yad Vashem' is a quote from Isaiah meaning 'a hand and a name'. Just one name. That we can comprehend.

My bar mitzvah was on a Saturday morning. Looking back on it now, the occasion was really for my parents, and as much as anything else, a celebration of them making a success of their lives in a new country, where they had arrived as refuges just over 20 years previously. My father by this time had developed a middle-aged spread, but was generally content having recovered from depression, and continued to liberally sprinkle his everyday language with short humorous anecdotes. My mother's good looks and trim figure were retained in her middle years. Her primary concern was to make her family happy, in particular her sons, to whom she was utterly devoted. The evening of the bar mitzvah was celebrated formally in central London at a venue that was highly fashionable at that time called Selby's.

Gerald Selby, the co-owner, had been cleared of performing a 'lewd, obscene and disgusting act' two years before when he was seen to kiss his fiancée on the lips in a car. The UK by the time of the trial had reached a watershed moment, where what had previously been seen as obscene was, by degrees, to become acceptable as norms for the age. The obscenity charge against Penguin Books for publishing the full unexpurgated version of Lady Chatterley's Lover *was similarly not upheld by the courts in England in 1959. Times were indeed changing! Bob Dylan described that theme in verse and tune a few short years later:*
Come gather 'round people

Wherever you roam
And admit that the waters
Around you have grown
And accept it that soon
You'll be drenched to the bone
If your time to you is worth savin'
And you better start swimmin'
Or you'll sink like a stone
For the times they are a-changin'

Our family's Selby's event was a dinner and dance hosting about 200 of my parents' friends. My friends were not invited – this was what was done or rather not done on such occasions in those long past days. My Auntie Betty from New York came over for the event. She gave me a clockwork powered 8 mm cine camera which was pretty much state of the art for the amateur in 1961. Each film came in a cassette which was loaded into the camera. When five minutes filming (no sound) was completed, I sent the cassette to Kodak for developing. Unlike the instantaneous viewing of videos on our smartphones nowadays, it took two weeks for the film to be processed and returned, when one needed a projector and screen to view it. But back to the celebration. I had to give a speech, which included me thanking my parents for bringing me up. It was recorded for posterity and is still in my possession on a vinyl record. Reverend Tachlisky encouraged me to speech very slowly 'like Winston Churchill' so that on the day I did not stammer.

In the mid 1960s my parents abandoned the coal, or should I say coke, fireplace in the living room. The hearth and accompanying metal mantlepiece above had likely had been installed when the house was built in the 1930s. When everyone had gone to bed and the lights had been switched off, the coke continued to burn blue in the centre and with yellow licking flames in the cooler periphery, which formed spectral shadows on the lounge ceiling. But truth be known it only heated one room. From time to time we put a paraffin heater in the other downstairs room. However the pungent fumes of combustion played havoc with my asthma. The bathroom had an electric bar heater fixed high on the wall. Dangerous if water splashed onto it, but a godsend in cold winters. In 1962 my parents had central heating installed. In my last years of school, our house welcomed friends of my brother and myself on Saturday afternoons to this now toasty warm semi. Every topic one could imagine was discussed – politics, science, religion and much more. My mother supplied copious amounts of home baked chocolate cake and freshly made coffee to assist and help prolong the debates.

Family summer holidays for about six consecutive years were in Switzerland. I have alluded to my first stay in Switzerland's lower mountain ranges in Lugano. I think my parents had a complex relationship with spoken German. On the one hand the Germanic Nazis had exterminated many of their family, and they had a reluctance to have anything to do with that country. Germany was strictly 'verboden'. No German cars or machinery. No articles of food. Anything with even a whiff of that country was totally out of bounds for them. On the other hand, German was their mother tongue, and continental Germanic Europe with its good coffee and language that they could freely express themselves in was an attractive proposition. That being so, holidays in Alpine German-speaking Switzerland was a compromise. We stayed in hotels in Alpine Arosa, Wengen and Grindelwald, which were accessed by rack-and-pinion railways. These railways interested my brother. I was more into the aircraft that took us to and from Switzerland. Formerly these were a turboprop Vickers Viscounts, as discussed above, and later on a de Havilland Comet 4B jet. The original Comets sustained a number of crashes due to fuselage weakness. However the cabin structure/fuselage was eventually made more robust, resulting in a safe and beautiful jet – the Comet 4B. It was operated by BEA – British European Airways – the forerunner of British Airways. Prior to one flight I had written to BEA to ask if I could look at the cockpit and was really thrilled when my name came over the Tannoy at the end of a flight requesting my presence with the captain!

A final word about my parents and Germany. Because of their ambivalence to speaking German, they often preferred to speak in English even to each other. My brother and I would have been able to speak German had they spoken it consistently at home. As it turned out we only learnt a smattering of everyday German. I learnt to mispronounce a number of English words that my parents taught me e.g. 'ben' for 'ban' and 'zilo' for 'silo', which was a source of mirth to my friends and an embarrassment to me.

The German government in the 1950s and 1960s was led by Konrad Adenauer who was the principal instigator of monetary restitution given to Jews who had lost property confiscated in the Nazi era. Both my parents received lifelong pensions from the German government. My father, who gave details of the family farm he had lost to the Nazis, informed me that embellishing the stated amount lost was both ethically and morally correct, considering the crimes that had been committed against him and his family.

And so to my sexual awakenings. The trickles of pre-adolescent sexual interest transmuted, day by day, to a flood, as each hair on my face appeared and with each lowering of the pitch of my voice. I could no more stem their flow than the proverbial Canute controlled the sea. Was I man enough, or rather adolescent enough, to ask a girl on date? The answer in my first teenage years was definitely NO! I was not good looking. I had a stammer. Any girl would laugh at me if I asked her out while contorting my face and spitting out the words. She would giggle and laugh at me along with her friends, just as they had done at my primary school. (In many senses dating online, almost universally de rigueur nowadays, would and likely does, bypass that perceived obstacle.) Anyway, what would I say to them? Not having a sister didn't help. I had no female friends. OK, maybe I could talk about planes, trains or science? Would they be interested in those topics? Likely not, my critical self answered. What about music? My tastes were definitely classical. I had a few 78 vinyl records of Beethoven and Mozart's music. They probably wouldn't be into that, but rather the pop scene of the early 1960s. Acker Bilk's 1962 hit *Stranger on the Shore* was the one, short, popular piece I could discuss. Discuss? What would I say? 'I like it'. That would sound lame. My thoughts of dating were filled with fear, rather than what I had heard other boys describe, which sounded like anticipation of sensual pleasure. How could I surmount these obstacles?

'OK, if one girl declines, ask another, and another. No, no that just would not do. I would become a laughing stock.' The ugly stammerer who asks every girl out, all of whom decline. 'Oh the shame!'

What would I do physically with them anyway if one did eventually say 'yes'? What was allowed? What would they want me to do? Kissing was probably OK, but breasts and the genital area were for married people. My knowledge of female anatomy and physiology was almost non existent. In spite of this ignorance (teaching myself O-level biology only empowered my knowledge of tadpoles, earthworms and plants!) somewhere deep in the recesses of my brain I knew more than my conscious mind let on. I had had a sexual dream about Celia Young when I was 11, during which I put my hand on her crotch (why would I know to put my hand there rather than her foot or the top of her head?) and then had a very pleasant feeling. I guess that must have been a dry 'wet' dream. While on the subject of dreams many of my nightmares as well as pleasant dreams prior to adolescence involved looking out of my bedroom window – either to see a witch zigzagging towards me – but never arriving – I woke up before that – or flying from my bedroom. Flying in a dream is a wonderful and exhilarating feeling – I have not dreamt it since childhood. I understand that one can have 'lucid' dreams whereby you can control what one dreams. Something to think about and maybe look forward to!

Back to the dating scene. When my brother was 17 he passed his driving test. Having access to a car was a distinct mark of virility within the circles I mixed. Akin to the peacock's multicoloured plume, or the alpha male gorilla beating his chest. Maybe I should wait till I was 17 and had passed the driving test?

My fear was oh so powerful. But in the end sexual drive and logic won the day. Approach had to overcome avoidance. After all I was 16, I was shaving, my voice had broken – there was no excuse for me not to make a move. I went with schoolfriends to a charity dance. The accepted modus vivendi at the dances was for the boy to ask the girl to dance. So with pounding heart and sweaty palms I asked, or rather shouted over the top of the blaring music, a girl with a pretty face and long blonde hair to dance. The act of shouting overcame my stammer. She was happy to go on the floor with me – the 'twist' was popular then although heaven knows I was rhythmically discoordinated (I think I still am – melodies OK, but rhythms … hm…). The booming music made it impossible to engage her in a meaningful conversation on the dancefloor. I motioned her to come outside. We chatted for a few minutes, mainly about the volume of the music and how she hated the whole process of meeting boys. We strolled along the outskirts of Hampstead Heath, stopping by a wooden bench. She smiled at me! I knew the kiss was imminent. Her lower lip was larger than the upper one, which had a quite pronounced cupid's bow. Would my kiss be a disaster? How should I position my head? Would our noses collide? Had I shaved adequately or was my lip hair still too bristly? Would she back off once our lips touched? In those few moments, between anticipating the kiss and lip contact, a myriad of these and other negative thoughts and visions raced through my mind. Cephalic collisions – like a car attempting to park in a narrow space, bumping into posts on either. But my head was on its way. It seemed to know what to do. I gently kissed her on the lips. They were soft and a touch sticky, likely from her lipstick. I went back for a little more. She seemed willing, and smiled at me between what became extended lip contacts. Hurray – a first kiss, brief, I had done it – well, one step of the way anyway. It was a rite of passage and I had succeeded. Such a shame that I did not have the presence of mind to savour the sensory moment in any greater detail. I did not see her again. Maybe we were kindred spirits, helping each other to negotiate the turbulence of growing up. A helping soul. Here and then gone.

First year university students in the USA who had never kissed a partner on the lips were compared to those who had kissed. The never kissed group were more likely to be Asian American, less likely to be in a romantic relationship, less extrovert, of higher academic achievement and drank less alcohol. Further detailed analysis showed the never kissed group were more neurotic, had mothers who were less facilitating for their offspring's independence and suffered lower self-esteem than those who had kissed.

The main factor defining a woman's attractiveness is thought to be her face, with the lips being a central feature of facial beauty. Ideally, the height (or width you might call it) of the upper lip should be less than that of the lower lip, and the upper lip should project 2 mm over the lower lip on profile. Pouty lips and a heavily defined Cupid's bow are considered to be sexually attractive. That said, beauty and facial attractiveness are hard to define. There have been attempts since antiquity to define and measure beauty numerically by use of 'the Golden ratio' represented by the Greek letter Phi, using two length a and b. Phi is obtained where a+b/a= a/b. Phi is then 1.61 or thereabouts. Phi or the 'Divine Proportion' is considered to be the key to aesthetic facial good looks. The width of the lips are Phi times the width of the nose. The length between the outer aspect of the eyes is Phi times the width of the lips. Leonardo da Vinci used Phi proportions in much of his art, including the Mona Lisa *and the* Last Supper.

Lefkowitz, ES, Wesche, R and Leavitt, CE. Never Been Kissed: Correlates of Lifetime Kissing Status in U.S. University Students. Archives of Sexual Behavior 2018 May;47(4):1283–1293.

Kar, M, Muluk, NB, Bafaqeeh, SA and Cingi, C. Is it possible to define the ideal lips? ACTA Otorhinolaryngologica Italica 2018 Feb;38(1):67–72.

When I was 17 I took the driving test. I had a number of lessons from a professional instructor, as well as time spent driving supervised by my father or brother. In 1965 hand signals were still necessary when taking the driving test. Contemporaneous hand signal with the right hand whilst changing gear with the left was an immediate fail, apart from being pretty dangerous. The hand signals to the modern driver look bizarre, more like saying hello to a friend across the street. Turning right was indicated by the right arm stretched straight out horizontally. Left turn by waving the right arm in a circle anticlockwise. Slowing down by waving the right hand up and down. There were more hand signals to learn if one approached a traffic policeman. At the end of the practical section of the test there was an oral theoretical test. I nearly came a cropper when asked to recall the stopping distances at various speeds. I knew the answer but found it difficult to enunciate. The examiner retorted, 'If you wait that long to respond you'll be dead'. I was again ashamed and angry that my stammer was letting me down. Nevertheless, he passed me. Happiness was having a full driving licence.

In the sixth form (year 12 and 13), I took three A-level subjects, chemistry, physics and zoology. I believe there were five of us who wanted to study medicine plus a handful of others who were going for dentistry who needed to pass the same subjects. Those studying medicine were David Lewis who became a haematologist, Peter Williams a GP, Jason Rapoport a renal physician, Pankaj Vadgama a professor of clinical biochemistry and myself. Another student was Chikwe Onyegbulum who had at some stage wanted to study medicine but died of Hodgkin's disease when we were all about 16. I can still hear and see him, emaciated from the cancer, saying how much he wanted to go to medical school. Today, with modern chemotherapy he would still be alive. This line is in his memory.

Our A-level chemistry teacher was a Mr Shale, who was in his late 60s, avuncular, jocular, cigarette always hanging from his lips but a good teacher and pleasant man with a pronounced West Country accent. He used the analogy of courting couples to explain chemical ionic bonding, where one element in a way donates an electron – so for example sodium chloride, where the sodium shares an electron with the chlorine atom, thereby converting two highly reactive elements into the very stable sodium chloride (salt). Mr Shale chuckled as he would tell us that the sodium loves the fluorine so much, it, or rather he, is happy to give away an electron and share it with the chlorine, which he construed as female. I have already described physics taught by Mr Rumble. In spite of his absent teaching everyone achieved an A grade – I guess we knew we had to get on with it on our own. A Mrs Webb taught zoology. She was rather stern, humourless woman who joined the department as we entered the sixth form. She was a poor teacher although her subject was difficult to impart. The essay questions at A-level were of university style and standard e.g. 'Compare the social life of insects to the social life of humans'. Not one of us got an A grade!

In the autumn of my last year I applied to medical school. I did not get many offers or interviews. One was at the London Hospital Medical College in Whitechapel. I believe my father helped me to secure an interview there. A Dr Gilliland attended my grandfather in his final illness, when he was impressed by my father's lifelong deep concern for his father. That being so, my father, I believe, asked Dr Gilliland, if he could help me get an interview. Viewed from well into the 21st century this seems totally nepotistic and unfairly influencing what should have been a level playing field. Maybe Dr Gilliland thought that the compassion and concern my father showed might be inherited via epigenetics or passed on to me via the nurture route. Be that as it may, I again found myself back in the East End of London in Whitechapel being interviewed by the Dean,

Dr Harry May. In his obituary it was said that he regarded selection of medical students as the most important task he undertook in his clinical and teaching practice. He believed in medical schools remaining small and disapproved of women taking up medicine. In the year I entered medical school, 1966, the London's entry was the youngest in the country with the smallest percentage of women – 7 per cent. Entry that year at the London included the highest proportion of working class entrants compared to other medical schools. Dr May's research interests included early work on antibiotics, a topic and speciality I would follow in years to come.

To return to the interview. My asthma was playing up that day, so I took a capsule, prescribed by my GP, containing theophylline, ephedrine and phenobarbitone. Looking back to that day, I believe that drug combination might well have helped to enhance my concentration and aided in the enunciation. But at the time I don't think I had made that pharmacological connection, merely thinking I had a good speech day by chance. I must have done well as I was given an acceptance on condition I obtained three grade Ds at A-level – not particularly high marks. Possibly if the Dean thought well of you, he would want to join them anyway whatever your grades? Doctors, I believe, do not need to be of the highest intelligence, but rather should have a hard work ethic and show empathy towards their fellow humans. I have heard it said that a medical student is a maths student who is not bright enough to follow that calling. However there are many types of 'intelligence' – for instance the ability to see the body in three dimension plus manual dexterity – those virtues making for an excellent surgeon.

So the last day of school arrived, we sang *Jerusalem*, I walked out and that was it – school years gone.

That summer in 1966 we watched with increasing hope and trepidation as England reached the final of the football World Cup. While the match was ongoing the streets were empty. We followed all of the final match against West Germany on BBC TV – in black and white. England's victory seemed to set the seal on me feeling upbeat about my future life. The day after the final, my brother, a friend, Jonny Webber and myself embarked on a two-week holiday in Scotland, taking it in turns to drive my father's white Austin A60 north of the border.

Details of that make and model for the aficionados are that it was a four door saloon powered by a 1.6 litre petrol engine, a top speed of 76 mph but would only cruise happily at 65 mph, took 24 seconds to reach 60 mph and cost just under £1,000 with tax. We put the car on the Motorail train in Kensington Olympia and travelled overnight by rail, to wake up and breakfast on the train that had reached and terminated at Perth Scotland. There was a distinctly romantic tinge in going to sleep in a dirty city and waking up in the foothills of Scotland's Highlands. These Motorail routes, mostly from London to Scotland or London to the West Country, were popular from the 1960s up until the end of the 1980s. After that time, with the advent of more powerful and reliable cars and fast motorway routes, the service was no longer profitable.

We toured the Highlands, the west coast, the picturesque fishing villages of Mallaig and Ullapool – and went further north to Nairn. I spent some hours taking to a hotel receptionist there – an attractive girl in her 20s who was dating an RAF officer at nearby Lossiemouth airbase. I don't think all could have been well with her relationship as she hinted quite strongly that she was free in two nights' time. But we had planned to travel on by that stage. Maybe a missed opportunity! I still have silent cine film of that holiday, us, the car and the train. Fixed and preserved in our prime on Kodak celluloid and now finally residing and resting in the hard drive of my desktop computer.

The end of August arrived, the results letter came through the door. I had achieved an A grade in chemistry, an A in physics and a D in zoology. The zoology was not a great surprise. So there I was – all set to start medical school in September 1966. The headteacher sent a note of congratulation with the result. I still have that note. Although medicine was my chosen course I had and still have a hankering to take physics further than A-level. This included astronomy, cosmology and Einstein's theories of relativity – both the special and general theories. However my maths was not up to university standard. That being the case physics was and remains an amateur fascination rather than a career.

Chapter 5

MEDICAL COLLEGE

And so I started my five-year study of medicine at The London Hospital Medical College in London's East End. The medical school was the first in the UK to obtain a charter to officially teach medicine. This was in 1785, soon after the first section of the hospital was built, just over a mile east of the Tower of London. The stated aim of setting up the hospital, a charitable institution, was 'The relief of all sick and diseased persons and, in particular, manufacturers, seamen in the merchant service and their wives and children in the area round the hospital'. The East End of London and its dockland had grown considerably in the latter part of the 18th century. A less charitable aspect of the hospital and its college was found in the excavations for the new Royal London Hospital in the early 21st century, where a forgotten cemetery was unearthed. Investigations exposed 262 burials containing a confusing mix of bones with extensive evidence of cadaveric dissection. Subsequent research by osteologists and historians revealed that in the early 19th century the hospital used their unclaimed deceased patients for dissection practice.

Whitechapel was also the home of the Whitechapel Bell Foundry which over the years constructed the bells of Big Ben and the Liberty Bell in Philadelphia. The area was also the home of Daniel Mendoza the English Prize fighting champion of the late 18th century, who was also the first Jew to be introduced to George III.

Back to the mid 20th century. In September 1966 I alighted from the East End section of the Metropolitan line (that section only was renamed the Hammersmith and City line in 1990). The passageway from trackside to exit was covered by a shabby metal and glass roof. The station has now been thoroughly modernised to make way for the deep Elizabeth line station. Exiting the old station, I was accosted by the sights and smells of the 'Chapel' market. Here was a fruit stall, displaying Golden Delicious apples, brown blushed yellow green Comice pears, deep red Morello cherries and bulging clusters of green grapes. And over there a hardware salesman displaying shiny aluminium pots of various sizes and metal hot water urns. Yet another vendor was selling bales of cloth – rusty brown heavy woollen weaves, yellow silken and multiple shades of green linen. In those days the stallholders were mainly East End cockneys, with a smattering of Bangladeshi men. The market is still there today, the vendors almost all being cockney-accented men from a Bangladeshi background. From the smells and sounds of the market I was confronted by the rising dark grey hospital exterior of the hospital, Neo-Roman style, with false inlaid Corinthian columns topped by a triangular pediment and with a base of five roman arches. This exterior is nowadays only a façade. It has been reborn as the 'Tower Hamlets Town Hall' behind which rises the new multistorey 'Royal London Hospital'. But in 1966 the old hospital still was very much alive. I walked past it and turned left into Turner Street, soon arriving at the medical school building – a partially redbrick box-like structure erected in the mid 1800s. The building is now used for recreational purposes. The preclinical teaching now takes place at Queen Mary College at Mile End.

I entered the building and went straight to the Anatomy lecture theatre on the ground floor as instructed. I believe only 95 turned up on that day! Andrew Brain, a senior medical student, instructed us on the dos and don'ts in rather officious and pompous manner; in particular what manner of attire we should daily present ourselves. Remember this was the 'swinging 60s'. I think no one took a blind bit of notice of his exhortations. We were split into groups of a four or five and told to go to the dissecting rooms on the top floor. There I was accosted by the powerful odour of formalin and 20 bodies underneath white sheets. The peak of each white mountain sheet was the underlying nose. The technician who undertook the body preservation had a sullen and to my mind, a ghoulish appearance. He set up the drips of preservative running into the veins of the newly arrived bodies. Although there were about five of us to a body, only one person could dissect, with another giving instructions from Solly Zuckerman's illustrated dissection manual. The remainder of the quintet looked on, chatted or combated the boredom by doing the *Times* Crossword. If a group had a problem, it could call on one of the 'demonstrators', who were qualified doctors wanting to take Part One of the Royal College of Surgeons diploma of FRCS, which necessitated an in-depth knowledge of anatomy. We were regularly tested on the relevant parts of the body, either being asked questions on skeletal anatomical landmarks or on cadaveric topography highlighted by small, coloured paper flags on pins inserted into the relevant part of the body – hence the tests were called 'flag days'. To help learn skeletal landmarks we each bought a half skeleton, which consisted of a whole skull and pelvis, one of each long bones (arms and legs) and vertebrae strung together. I bought a set in the medical college bookshop. They were real bones. Osseous remains of a person who had once walked the Earth! However we did not know the provenance of what we purchased, but were likely bony remains from the Far East, where the owner had been paid in life for the ultimate use of their body. I found the dissecting experience not surprisingly morbid, the fullest sense of that word. Our dissection was frequently rather amateurish, and did not clearly demonstrate the anatomy. My body was an elderly looking man, whose name was written in indelible black ink on various points of his torso. When we had finished our dissection at the end of 18 months, his various limbs and organs were interred in a coffin and presumably sent for some sort of burial. Textbooks such as *Gray's Anatomy*, and shorter more pictorial atlases helped us understand the three-dimensional structure by means of clear two-dimensional illustrations. However there was no logic in either the names, say of muscles, or where they were in relation to other structures. A muscle is attached at least at two points, commonly termed its insertion and origin. The exact location of these sites had to be learned by rote, which was not easy for me. The occasional muscle name made sense, such as the sartorius, which crosses the front of the thigh. Sartor is Latin for tailor. Seemingly tailors in Roman times sat cross legged as they worked. Nowadays learning anatomy is easier, using 3D computer modelling or observed much more clearly at live surgical operation. We also studied histology and embryology, which are the microstructure of various organs and their formation in the foetus respectively. The medical school museum contained the macabre pickled remains of malformed infants which were displayed for teaching.

The anatomy department was headed by Professor Richard J Harrison, who had made a name for himself by researching the comparative reproductive anatomy of various mammals, and went on to undertake physiological research relating to diving mammals. His diction at lectures was rapid and rather mumbling, and he seemed to over salivate as if he words tumbled out too quickly for him to even swallow. He later went on to a prestigious chair in Cambridge, as well as a knighthood.

Physiology was an extremely important subject in the preclinical course, which lasted for just over 18 months. On our first day of physiology, amongst other procedures, we took each other's blood pressure, with an old style mercury sphygmomanometer. Tim Cox who I was partnered with, told me mine was high – something like 180/100 and that I should get treatment for it. My labile blood pressure has haunted me for all of my life, and is a reflection of my anxiety which then – doing something strange and new like BP assessment, tended to cause a surge in fear. Tim went on to become a professor of medicine in Cambridge. Back to 1966 – we then took blood from each other. Just a brief lecture on how to do it – and off we went! Quite a few of the class fainted! Professor Kenneth Cross was in charge of the physiology department. His research interest was neonatal physiology, so that he was as much a physician as a physiologist. He also happened to be my personal tutor and was a kind man. At the time of the 1967 Six-Day Arab–Israeli War he took the trouble to ask me if I had any relatives in Israel and how was I feeling about the situation. I only found out later that he was a lifelong depressive, as was his first wife.

The third major arm of the pre-clinical course – also known as the second MB (the 3rd MB was the final examination) was clinical biochemistry. Viewed from the present day, many of the practical biochemistry sessions we undertook seem rather pointless. For instance assaying the blood levels of haemoglobin – currently as a doctor one just requests it from the lab – I do not believe we really needed to know the exact biochemical method for so doing. Another practical involved us measuring the level of glucose in our urine. This is now done using dipstix – obtaining the result in seconds by the bedside or outpatient clinic. We had to learn the Krebs cycle off by heart – that is how sugars yield energy within cells. The exact pathway involves multiple steps – I memorised it for the day of the test. Again it was necessary to commit this to memory – to be regurgitated in the exam room and then forgotten. All this criticism said, the staff of the biochemistry unit were pleasant and helpful. Professor Warren led the unit. He was a diminutive, slim man with a bushy moustache. Practicals were set up by an obviously camp gay man who generously assisted us with the experiments.

The social life the Medical College involved hard drinking by many of the men. I did not get too involved with this. Didn't see the point as alcohol served only to render me more tacit and dull witted than when sober. As if to balance the paucity of female medical students, there were many attractive female student nurses, physiotherapists and dieticians on campus. I dated a student nurse for three months in the first year. I met her at a medical college 'hop' (a word no longer in current usage alas, meaning an informal dance). She was talking to a larger version of herself, who turned out to be her sister. Time dims the memory, and I have no photos of her. She was of medium height, had a trim figure, had long, dark blonde hair and spoke in a croaky Glaswegian accent. A minor blemish to this Scottish goddess was a large gap between her upper incisors. She transferred her training to Glasgow to be nearer her mother who was not well. I wonder where she is now?

I remember
The smell of old wood and floorboards
Waiting for you
And your cloak that was blackly
Rough on the outside
And red and silky smooth within
And how it swirled and enveloped me
And you smelled clean and Amplexy
Soft and young
With long fragrant hair

And your passion.
I have to remember
Because you are no longer there
And the buildings, cloak and vapours
Just memories that visit now and then
I am telling me and you this
If you are still anywhere.

The second MB written exams were held after 18 months. Below are a selection of some of the questions.

'What does the tibialis anterior muscle do in everyday life?' (lifts foot up and everts it)

'Discuss the term teratogenicity' (the time of the exam was just a few years after the breaking of the thalidomide disaster, which affected about 10,000 babies, causing among other problems phocomelia i.e. severe limb deformities)

'How is the water content of the body regulated?'

'How does the body adjust to and after exercise?'

'What is the neural mechanism involved in drawing a straight line?'

'What are immediate and long term effects of altitude on the body?'

There were also a number of talks on medical statistics – lectures given by a pharmacologist Dr CRB Joyce. He later went on to work for the pharmaceutical company Ciba-Geigy in Switzerland. The statistics exam had four questions – one had to apply statistical analysis to given data, to assess their significance. In those days (and maybe still now) the examiners used a 'close' marking system which ranged from a fail which was under 50 per cent to top marks which were 54 per cent. This was in all subjects.

As medical college years went by I got to know a few of my fellow students. Hugo came from Hastings, liked strongly spiced Indian food and had a large shaggy beard. He became an oncologist. Chris Roberts was a quietly spoken musician who became a GP in London. John Wilson played the bagpipes and eventually emigrated to Canada.

In the summer of 1967 I drove to Malvern and its surrounding hills with a few friends. My brother and his girlfriend Valerie were also there. They were both studying law. Valerie brought along her younger sister Tricia. My memory of that event is of a warm, glowing day and the steeply inclined rolling green hills, from which we all gazed at the wonderful vistas of Herefordshire and Worcestershire below. I was very conscious at the time of being in Elgar country. He likely walked and flew kites from the very same uplands that we walked through. Ever since then, whenever I hear the Enigma *Variations*, feelings of spiritual warmth and hope mixed with the memory of those times wash over me.

I chatted with Tricia as we rambled over the hills. What were my thoughts and feelings at that time on the girl who became the love of my life and the bearer of my children? I use the world 'girl', as that is what she was at that time, studying for A-levels at a good direct grant school in Birmingham. To my mind this young woman was physical perfection. Her dark hair was cut short and layered with a bob or fullness at the back. She had brown almond shaped eyes that curved at their outer edges in mild pseudo Asian configuration. I felt more than a touch nervous talking to her as we rambled over the hills. She walked with a bouncy stride – there was something in that gait that I also found attractive, although it was a far cry from the gliding deportment that young ladies of the age were taught at Lucy Clayton's 'finishing school'. Something we said seemed to amuse her, at which she broke out in an enormous smile, which almost floored me with desire. She was wore brown suede trousers and informed me with some enthusiasm that she was dating a boy called 'Vamp'. I did not think there was any point in me asking to see her again. However, I met her again in the winter holidays of 1967, that was about five months later. She was no longer seeing 'Vamp', at least not regularly. I wrote a letter to her immediately after her return to Birmingham saying how much I liked her. I was delighted when she wrote back to say she felt the same. We wrote to each other for the next two years, with me based in London and Tricia at first finishing school in Birmingham, and later on spending nine months in Manchester studying psychology. These 60 letters are still in our possession well over 50 years later. They contain events and ideas of that era as well as being love letters.

Some of the dates and letter contents are as follows:

February 1968. From David to Tricia (D to T). An old friend of my family, an Edwardian lady I called Auntie Flo, told me that I did not have a stammer till I was 3 and that I began this speech defect to take my parents' attention way from my brother.

March 1968. D to T. I saw the anti-Vietnam war rioting outside the US Embassy in Grosvenor Square London on the TV. Hundreds of protesters and police were injured.

April 1968. D to T. My father told me that good old wines are often covered in dust and cobwebs.

May 1968. D to T. went to a late night movie with my friend Irving. Saw 'Reflections in a Golden Eye' with Marlon Brando and Elizabeth Taylor – a story of sex and betrayal at a military base. After leaving the cinema we had to work our way through crowds of drunken West Bromwich Albion fans.

May 1968 D to T. My parents' friends, the Chippecks, had opened up a dance club named The Purple Pussycat in the Finchley Road. One of the owners' sons Ricky was charged with actual body harm committed on an Israeli patron.

September 1968. T to D. Tricia's Uncle Bill who was a GP in Manchester where she was studying told her that he never gives the oral contraceptive pill to unmarried women!

October 1968. D to T. Alerting T that he had pubic crabs, and that both T and D should eradicate them with benzyl benzoate lotion.

Early December 1968. D to T. UK Cardinal Heenan was reported on the 'Frost Report' (a satirical TV programme) saying the oral contraceptive pill was a physical danger to women. Later watched Dr Who starring Patrick Troughton, who was fighting the alien Cybermen.

Late December 1968. D to T. A senior surgical registrar discussed intestinal obstruction, which is usually caused by bowel cancer. A case he had seen 2 days before our teaching session was unusual. He had retrieved a condom loaded with faeces that had caused the problem. The shamefaced patient told him that a husband had returned home early and found his wife in flagrante delicto with himself. The husband gave the patient a choice – a beating or swallowing a condom. He chose the latter.

Early June 1969. D to T. Helped Dr Donald Scott (a neurophysiologist) do research work on the use of stroboscopes in the diagnosis of epilepsy. Some of us medical students volunteered to be subjects and were attached to multiple electroencephalogram (EEG) leads. Stroboscope flashing lights can induce epileptic fits in those who are vulnerable. None of the students including myself had a fit. Lucky for us and Dr Scott!

Early June 1969 D to T. D upset with his father who called D's Mum a fool. She became upset, which induced an asthma attack in her. She told D's father he should have married someone else, and that she had no choice and had to marry him. She reminded him that he had begged her on bended knees to marry him, and that at first she declined because she didn't really like him. D felt upset and confused. Later that evening D's father slapped D and his brother.

July 1969 T to D. A psychologist on the TV reported that air travel takes you back to childhood because you are 'tucked in your seat by the air stewardess' in the same way as a Mum tucked you in bed as a child!

October 1969 D to T. A surgeon gave a talk on stomach ulcers. He had a deep voice and wore and a three piece suit with a red speckled bow tie. He explained that ulcers were formed where 'aggression' outweighed the body's 'defences'. Aggression manifested as excess stomach acid. Defences were the mucus the stomach produced. He hit a blackboard behind him with his fist, exclaiming 'we must cut aggression'. It would be another 12 years before ranitidine was introduced, which drastically cut the formation of stomach acid and made surgical operations for peptic ulcers almost redundant.

November 1969 D to T. Went to see George Chapman today to see if he can help my stammer. He is a 'medium' and spirit healer and conduit who performed spiritual surgical operations undertaken by Mr William Lang, who was an ophthalmic surgeon practicing at Moorfields Eye Hospital until his death in 1937. Chapman's house is called St Brides and is on a main road with a quaint front door knocker and front doorbell that played a tune. A middle aged woman opened the door and asked me my name. As I waited to see the healer I notice his consulting room had the name of the deceased 'Mr Lang' on the door rather than Mr Chapman. The woman told me that some very famous people had visited him. While I was there a vicar brought his wife for healing. The woman told the vicar that Mr Chapman was in a trance for up to 6 hours a day. My session cost me £2 and 2 shillings *(equivalent today is £110). I was called in to see George Chapman. The room was dark, with venetian blinds down and curtains drawn shut. The woman, who had introduced herself as his secretary, introduced me to Mr Chapman who looked about aged 50. His eyes were tightly screwed up. He had long greying hair and stood bent as an old man. He extended his hand to me, saying 'Ah Mr Goldmeier, pray you be seated' in a croaky voice. His hand grip was firm, very firm. I did not need to tell him why I had come, although I had not spoken so far. He chatted to me about The London hospital as it was in 1870, and how he enjoyed seeing Jewish patients in London's East End, adding that his father's name had been Israel and his grandfather's Isaac. As he spoke he grimaced, showing his teeth. He then took me round his room with his hand on my shoulder and showed me pictures of himself in life as well as of his son Basil who had also been a surgeon. He told me some of the great current medical practitioners who had consulted him as well as a heart transplant surgeon who I took to be Christian Barnard. He said he did not agree with some of the ethics of organ transplantation. Then he said, 'Sit you down on the couch', after which he told me he was going to operate on my spirit vocal cord and salivary glands. He continuously flicked his fingers and asked Basil to put in stitches here and there. All this did not hurt me. When he finished he said he could not promise success but that I was 'healable' in time. I was instructed not to talk for a few hours, not to eat till the next day and to update my progress by letter in a month.*

Other memorable events discussed in the letters included the release of the Beatles' album Sgt Pepper's, the Torrey Canyon oil spill and of course the 1967 Middle East war.

I put a great deal of effort and time into revising for the 2nd MB examination and being my usual despondent self, thought I would fail. Tricia gave me psychological support from afar. I passed. I had wanted to study for a further 18 months undertaking a BSc in physiology prior to working on the wards. However my marks were not good enough for that. I think my anxiety and consequent suboptimal concentration levels were largely to blame. Everyone's exam results were posted for all to see outside the medical school library. A word about the librarian, a Mr Entract. You may say, 'what an odd name', the word entr'acte meaning music or some sort of performance between acts of a play. Well, the librarian lived up to his musical name. He sat in his open office, silver haired, in a long white coat (never before or since seen a librarian in a white coat), squeezing and stretching his brass covered concertina in a succession of major keys, his fingers dancing an Irish jig over the buttons, while accompanying himself in low volume bass Gregorian-type chant. His entr'acte could go on for an hour or two, disturbing those wanting to use the library for reading or revising. He was a bully of a man, but like so many of that ilk would cease his disharmonious tumult if someone told him in firm terms to shut up.

So I became a clinical medical student. We wore long white coats with a horizontal dark blue stripe on the upper left pocket – to distinguish us from the all-white coated qualified doctors. The consultants however did not wear white coats, rather suits were de rigueur for these senior staff. This custom lasted well into the 2000s in the UK, after which short sleeved shirts with open neck and no tie and indeed no white coat became the norm for all staff not wearing uniforms or scrubs. This was in response to rising nosocomial infection (meaning infection acquired in hospital) rates and the insight that clothing such as ties could transmit pathogens.

Three London Hospital nurses in their uniforms

Courtesy of Barts Health NHS Trust Archives

The London hospital nurses uniform in the 1960s was lilac for staff nurses and deep blue for ward sisters. The shoulders had puffed 'leg of mutton' shouldered sleeves. Aprons and caps were of starched white linen.

A group of about five medical students was attached to each 'firm', which usually consisted of two consultants, a senior registrar and/or a more junior registrar and a newly qualified houseman. The inpatient wards were within the east and west wings at four levels. As students we 'clerked' the patients. This meant introducing oneself to the patient, politely asking if it was OK to take a history and then examine them. One elderly inpatient who I clerked suffered from an exacerbation of chronic bronchitis. He was thin to the point of emaciation because of emphysema and struggled to voice me his name and when I asked him why he continued to smoke 60 a day. "ow else can I get my phlegm up in the mornings doc?' Another 80-year-old Jewish patient could not bring himself to directly say he was severely constipated. We were taught to listen to or 'auscultate' (from the Latin auscultare meaning to listen) the heart and lungs using a stethoscope. The best stethoscope was and I believe still is a Littmann. A stethoscope is highly symbolic of being part of the medical profession and we medical students displayed it proudly either protruding out of our white coat pockets or round our necks with one earpiece tubing either side of our necks. The fashion nowadays is to drape the whole apparatus round one's neck. Retiring from medical practice is termed 'hanging up one's stethoscope'.

Our histories and findings were written onto fawn coloured sheets, kept with the actual clinical notes in file, and presented verbally to the consultants or whoever undertook the teaching ward rounds. My stammer was continuing to cause problems for me and for the staff, who either appeared annoyed or embarrassed by my efforts and frequently cut my presentations short. The anticipation and fear of having to present at ward rounds seeped into my daytime thoughts and sleeping reveries. I decided it was again time to try to do something about it. My attempts at clear enunciation distracted me from forming a clear, logical content to what I had to say, so that the end product of the presentation was poor in both diction and substance. As I said, I am sure the consultants and teachers who listened to my exposition just wanted to pass on to other matters and other students.

I had heard that a GP in Mill Hill, a Dr David Waxman was a highly competent hypnotherapist, and in his writings had presented successful treatment of stammerers. Trance-like states had been reported in ancient Egyptian and Greek literature. Hypnosis developed into a medically respectable mode of treatment when used by Drs Breuer and Freud in late Victorian Vienna, although Freud himself deemed his treatment outcomes were poor, hence his foray into psychoanalysis. The most famous medical hypnotherapist of the 20th century was Milton Erickson. He is considered the father of modern hypnotherapy. He used indirect suggestion, metaphor and storytelling to alter behaviour, rather than direct suggestion. Most of us recognise a trance-like state when, for instance, we are waiting at a bus stop or on a train journey as the wheels sound out a regular rhythm. David Waxman himself was the founder of what in the 20th century was named the section of Medical and Dental Hypnosis of the Royal Society of Medicine. He had written a number of books on the topic. My father paid for five sessions with him. I saw him at his private practice in Mill Hill. I lay on his couch as he attempted to induce a trance. His voice was pleasant enough. But in terms of my stammer the sessions were totally useless. I was too tense to enter into a trance -like state. However I do remember him saying that, 'Medical students are the lowest form of animal life'. I didn't really need him to say that!

Surgery implied entering into a different world, with some time being spent in the operating theatres. Before setting foot in the operating area, one changed into scrubs and a mask. If one was lucky enough to assist then one washed and scrubbed up, followed by donning a sterile, dark green, surgical gown, with a nurse tying up the strings at the back. Lastly, putting on sterile rubber gloves without touching their outside working surface. Once 'sterile' it was vital not to contaminate the hands by touching anyone or anything non sterile. This meant positioning the hands together close to one's body while walking around the theatre. On my first day in theatres I accidentally brushed against a surgical registrar, who screamed, 'I am sterile'. I laughed inwardly at this double entendre!

A very senior surgeon, Mr ECB Butler was attempting to remove a benignly enlarged parotid salivary gland. Doing this can result in damaging the facial nerve, a branch of which runs through the gland. So he used an electrical stimulator, which he termed 'the twitcher', to test areas he thought might contain a nerve branch. If the face twitched where he considered incising, then he would steer clear of that route. He must have been at least 65 years old when I saw him at work, a tall, thin white-haired man who was taciturn as he worked. I was rather shocked by his very shaky hands – I wondered if he should be undertaking this surgical procedure, as it involved intricate and careful dissection. He certainly would not be allowed to do so nowadays! When not operating he was an excellent teacher. On one of his ward rounds he showed us a man with a black gangrenous toe. He pulled at it (the man could feel no pain so likely had a diabetic neuropathy) resulting in it coming off in his hand, exposing bone and raw flesh beneath. 'We'll need to tidy that up later,' he said as an aside to the ward sister. I later found out that he was in the twilight of a varied and fascinating career. His rather gloomy and aloof disposition was seemingly underpinned by shyness. He had been the ship's surgeon on the maiden voyage of the Queen Mary in May 1936, and had remained in that position for a further 30 transatlantic trips. In 1944 he was one of the first surgeons to successfully treat osteomyelitis, that is infection of bone, with penicillin. Prior to this osteomyelitis had a mortality rate of 25 per cent in adults and 80 per cent in children. Returning to the time I spent in theatres with him, at one point in an abdominal operation he was clearly having trouble, and became rather agitated. So Professor David Ritchie was summoned. He was a gifted polymath who held the chair in surgery, he completed the operation on Mr Butler's behalf. David Ritchie had many claims to fame, including the early use of hyperbaric oxygen to treat the Himalayan mountain climber Chris Bonington's frostbite.

While attached to a surgical 'firm' I went to the 'Receiving Room' (what we would now call Accident and Emergency) with the registrar to see cases of hitherto undiagnosed abdominal pain. An accurate description of the Receiving Room in the Edwardian era is depicted in the television miniseries *Casualty 1909* and is well worth viewing for those who wish to understand how emergency medicine functioned pre World War 1. Reverting back to me seeing patients with abdominal pain – this not uncommonly led to an appendicectomy. We were also given brief instruction in the techniques of surgical stitching and tying suture knots rapidly and efficiently, which meant we were then allowed to repair skin lacerations.

A pathology demonstration marked most weekday lunchtimes, taking place in the Institute of Pathology. The post mortem findings of patients who had recently died were demonstrated and presented along with their in life clinical details. Students stood at the rear of a small but up to date modern tiered theatre, while one of the demonstrators, often John Summerfield, who went on to be professor of medicine at Imperial College, held up the naked eye findings. The dissected bodies lay on large metal tables. The sights and odours of the newly cut up bodies did not entirely help me to digest my lunch!

The academic discussions and debate of these sessions were led by Professor Israel Doniach, who was a small, slim, bald headed, quietly spoken man with an encyclopaedic knowledge of medicine and pathology. He was often accompanied by Professor Henry Urich, who had trained in Poland and who was an expert on neuropathology.

Once every 2–3 weeks there was a clinicopathological conference in the steeply tiered all wooden Bearsted lecture theatre. A presenter gave a few relevant clinical details, not enough for the average doctor to reach a diagnosis. However for the highly experienced physicians who reached a diagnosis with impeccable logic, it seemed all too easy. Their clinical diagnosis was confirmed by histological diagnosis which was presented after the clinical diagnosis had been elucidated. Frank Goodwin, a nephrologist and general physician, with his slim, taut figure and flamboyant large bow tie, was one such discussant who gave sparkling displays of logic and medical erudition.

After the first year of basic medicine and surgery we moved to specialities and spent time away from the London at other peripheral hospitals in Essex.

Psychiatry was undertaken at both the London Hospital and St Clement's Hospital in the Mile End Road. St Clement's was built in 1849 as the City of London workhouse and later became an independent psychiatric hospital. It was heavily bombed in World War 2. In the new millennium it was replaced by up to date housing. At St Clement's I sat in on ward rounds undertaken by Dr John Denham who had fled Nazi-occupied Austria just as he was about to complete his medical studies. He had swept back greying hair and an attractive smile and personality. I have two strong memories of him. One was that when discussing patients he seemed to be addressing and talking only to an attractive psychiatric social worker rather than anyone else present. One of the patients, an angry paranoid schizophrenic sparred verbally with him. After a few minutes the patient picked up the rather large and heavy desk separating him from Dr Denham and threatened to throw it at him. Dr Denham replied in a very calm voice, 'Just put that down', which the patient reluctantly undertook. All the while Dr Denham was smiling at the social worker. The patient then went onto say, 'You're nothing but an Austrian Jew'. Dr Denham did not seem to appreciate this comment, and became visibly angry. At that point he instructed the nurses to give the patient more chlorpromazine (a major tranquilliser)!

Dr Sidney Crown was a psychiatrist who worked solely at the London, which had a small inpatient unit. He had formerly been a clinical psychologist. Fast forwarding a few years to when I was qualified, I rather desperately asked if I could see him to discuss my anxiety, mainly in relation to my stammer. He was very sympathetic and suggested I join the Group Analytical Practice in Baker Street. I went there twice a week for 18 months. Most of the other attendees were either doctors or psychologists. The Group leader was a Jim Hume, an older man who said very little. I derived no useful feedback from either him or the participants in regard to my stammer or anxiety. At the time I was doing research into the cause of rectal inflammation in gay men. The participants felt I had picked the topic because I was a latent homosexual. Instead of just letting their words wash over me I protested, which just confirmed these suspicions in their minds. I cannot recall anything that the group leader said except, as if rising from deep slumber (maybe he was asleep!) to utter the great words of wisdom 'I feel the group is saying it doesn't love me'. It is worth saying that my NHS bosses at the time were quite content about me taking time out from work twice a week to attend the group.

Back to medical student days. Obstetrics and Gynaecology was done both at the London and a peripheral hospital – in my case Billericay, where the registrar who taught us was one Bob Winston who went on to become Lord Winston of infertility fame. We medical students were instructed by lecture or tutorial how to repair a lower vaginal birth tear. We then saw one being undertaken by a senior doctor, after which we undertook the repairs on our own without supervision. I dread to think the awful mistakes that were made by so doing. I know of a woman attended to by a student in my year who came back to the hospital some weeks after the repair to angrily report it was impossible to have intercourse because her opening was too narrow – an overenthusiastic medical student having almost closed her vagina with his sewing handiwork!

The final stage of labour is never a dull event. Verbal altercations between women in the throes of labour pains and their midwife was not uncommon.

'I don't want this f***ing baby no more' to which the midwife would reply, 'You should have thought about that nine months ago mother'. One of the maternity wards at the London was named Marie Celeste. The etymology of the name is uncertain. It is possibly the Marie Celeste was Galileo's daughter who took religious orders, or the 18th-century saint of that name or even a number of more modern day women who took holy orders. Whosoever the ward was named after remains a mystery, much as the fate of the similarly named ghost ship Mary (not Marie) Celeste. However, there was a ghost connection between ship and ward. It was said that the ghost of Marie Celeste, whoever she was, haunted the ward and would appear to dying women, offering them a cup of tea!

We spent a few days working 'on the district'. That meant going out to women who were having home births around the Whitechapel area, very much as described in *Call the Midwife*, and indeed in an almost identical locality and only a handful of years later. However there were no district midwives who were nuns that I ever came across. The senior district midwife at the time was an older grey-haired woman who we all called 'Auntie Gladys'. I have no doubt looking back that she was both highly experienced and dedicated to her vocation. Nevertheless she seemed ancient to us at the time, exuding the scent of the art of midwifery rather than the dashing new science. No one ever mentioned her real name.

She rode her ancient pushbike with great determination and at a ferocious pace which belied her years; a veritable accoucheuse, Wizard of Oz, Witch of the West hybrid! The medical student followed on their bike, strangely enough trying to keep up with her savage pace. Both bikes had large wicker baskets into which were placed the needful for home deliveries in battered leather suitcases. The homes around Whitechapel at that time were often dilapidated and damp – exactly as described in *Call the Midwife*. Once we had arrived at the mother's home, Auntie Gladys gave 'doctor' instruction after instruction. 'Doctor will not put the cases on the floor' (that appeared to have been a grievous sin), 'Doctor will unwrap the sterile pack', 'Doctor will wash his hands' and so on. One of my Welsh student colleagues was less polite than I, and told Auntie Gladys 'Doctor will not do that. Doctor will now go back to have a pint of beer, Doctor says goodbye'. A first time mother expressed consternation when seeing her newborn covered in a white cheesy material. 'Wasat white gunk on the littlun, it don't look right to me?'. Auntie Gladys replied in a calm, measured tone, 'That's vernix my dear. It's nature's cold cream to protect your baby'. Vernix, which is composed of skin and fatty cells is a creamy biofilm that protects the newborn and facilitates extra uterine adaption during the first postnatal week. It is often wiped off so as to present a pristine baby to the world. However more recently it has come to be understood that it helps the newborn to control their temperature and enhances their innate immunity. Auntie Gladys did not know the science behind the benefits of vernix but her immense clinical experience told her vernix was there for a reason and should not be summarily wiped off.

Singh, G and Archana, G. Unravelling the mystery of vernix caseosa. *Indian Journal of Dermatology* 2008;53(2):54–60.

Back to medicine at The London. I presented a case to Dr Ronny Henson the senior neurologist. Dr Henson's voice had an almost musical timbre, which likely reflected his love of music, via which he had met his wife. The woman who I discussed that day had a meningioma, that is a benign tumour of the brain coverings. However, in her case the tumour was large and because it pressed on her brain she presented with neuropsychiatric symptoms such that what she said was rather humorous but also illogical. Unfortunately my stammer got the worse of me on that day. I recall trying to circumlocute words that were difficult for me to enunciate and used the phrase 'left hand side' rather than 'left'. This seemed to annoy Dr Henson. By the end of my presentation I felt I had been totally taken to pieces and pretty useless. The patient underwent surgery but died of haemorrhage on the operating table.

I spent four weeks at St Margaret's hospital Epping, seeing a variety of patients. I will discuss more of that hospital further on, as I did two jobs there after qualifying. While there, in 1969, I saw Concorde roar over the hospital at low level – it must have been one of its earliest UK flights and was a sight and sound one could not easily forget.

By this time I was spending more time with Tricia. She had left university in Manchester to be with me and had been accepted to study psychology at Bedford College in London, where she qualified with First Class honours and a prestigious prize, being the top in her year. In the six months before she started her London course she worked in Carnaby Street in a clothes shop called Lady Jane. This was owned by Harry Fox who in trying to advertise his shop, was arrested for causing an obstruction by having live models change clothes in the shop window. Carnaby Street was a focal point in the burgeoning fame of that mecca of 1960s fashion, with its vibrant clothes, often with clashing colours, its new cultures, exciting new music and general rebelliousness. There was unfortunately a dark side to that era. More than one of Tricia's work colleagues died of a drug overdose during the time she worked there. Many of the sales girls working with Tricia wore bizarre dresses, with bead necklaces, large earrings and bare feet completing the anarchistic picture. I asked one of Tricia's workmates why she wore no shoes. ''Cause it's sexy', was reply. Another told me it was the way to dress in those times.

FOOT FETISH – CASE HISTORY CIRCA 2005

Foot fetishism did not present frequently to my clinic.

The man was a 50-year-old Roman Catholic priest who worked as botany lecturer at a well-known Italian university. His family were very proud that he had taken holy orders. About five years before coming to see me he had wanted to marry the woman he had fallen in love with. The marriage took place in secret in Italy, with no members of either of their families being present. The couple immediately flew to the UK, hoping that no one would be able to trace them. He made a living by teaching biology A-level at a private school. His wife was 35 years old and not in the best of health as she had an anaemia of uncertain cause. They were keen to have children. However he could not obtain an erection except by touching, massaging and fondling her feet. In our sessions he did most of the talking, in large part because his English was good and hers almost non existent. There was nothing obvious in his past life to account for his foot fetish. My psychologist colleague and I devised a behavioural programme whereby he took sildenafil (Viagra) before sex, massaged her feet, and while the erection persisted transferred himself and his erect penis to lie on top of her. As time went on he became accustomed to assuming the missionary position with an erect penis and was later able to place the penis on the vulva and eventually inside the vagina ending. Ejaculation inside the vagina necessitated him thinking about her feet ... As time went on he no longer needed to use her feet as a sexual initiator.

The word 'fetish' is usually used to describe an inanimate object which can arouse sexual desire, but may also apply to parts of the body. In 2015 a man was arrested for having sex with 450 tractors, with a special penchant for Massey Ferguson makes. According to DSM 5, a paraphilic focus of a fetish disorder involves the repetitive use of an object or body part (typically non genital) to achieve sexual arousal. It is unclear exactly how common fetishes are in the community. Scorolli and co-investigators used the internet to examine 381 respondents in discussion groups relating to fetishism. About a third of the individuals targeted had a preference for body parts. Feet and toes were the commonest reported body parts and were five times commoner than the next preference (for body fluids). The authors accept this was a biased sample.

There are a number of theories as to the cause of fetishism, including psychoanalytical, where Freud suggested feet were a penis substitute consequent on childhood castration complex. Behavioural factors might be that men could learn to become sexually aroused in the absence of classical erotic stimuli when they were provided with appropriate feedback. It has been proposed that because the region in the brain that processes sensory input for the feet is in close proximity to the genital representation, that an accidental anatomical and physiological overlap takes place. It may have been that our patient's cultural and religious values and behaviours focused his sexual interests away from the genitals onto a region more acceptable to his values.

Giannini and colleagues have pointed out that each of three epidemics of sexually transmitted diseases in the previous millennium was associated with a sexual focus on the female foot. These disappeared after the epidemics ceased. During non-epidemic epochs eroticism was attached to breasts, buttocks and thighs. The 13th century 'gonorrhoea' epidemic was associated with female feet being eulogised in Romance of the Rose and by other poets. The epidemic of 16th century syphilis coincided with artists specialising in depicting the female foot with the great toe and next toe forming the 'toe cleavage'. A further syphilis epidemic just prior to the first antimicrobial Salvarsan 606, resulted in brothels specialising in foot fetishes. English dairy shops at the time sold 'Trilbys', a vanilla ice cream in the shape of a foot. During the more recent HIV epidemic, feet were considered 'safe', particularly for gay men whilst there was no effective treatment. This of course changed with the advent of effective HIV treatment in 1995.

Scorolli, C, Ghirlanda, S, Enquist, M, Zattoni, S and Jannini, EA. Relative prevalence of different fetishes. International Journal of Impotence Research 2007 Jul–Aug;19(4):432–7.

Giannini, AJ, Colapietro, G, Slaby, AE, Melemis, SM and Bowman, RK. Sexualization of the female foot as a response to sexually transmitted epidemics: a preliminary study. Psychological Reports 1998 Oct;83(2):491–8.

Ventriglio, A, Bhat, PS, Torales, J and Bhugra, D. Sexuality in the 21st century: Leather or rubber? Fetishism explained. Medical Journal Armed Forces India 2019 Apr;75(2):121–4.

Tricia really enjoyed her six months working in Carnaby Street and was an excellent salesperson, being able to sell clothes to almost anyone, including women of larger proportions who didn't really fit into the smaller sized clothes that were available on sale. These were typically over-enthusiastic middle aged Americans who had more money than sense and who had been caught up in the frenzy of 1960s fashion! I frequently stayed overnight with Tricia at her grandmother's (who had passed away some years before) house in Hurlingham near Putney. She came with me to see the film *Far From the Madding Crowd* which had a showing at the Medical College. The students' union had to hire a 16 mm film projector for the occasion.

In 1969 I went on a student tour of Israel with Tricia for four weeks. There was a freshness and self-confident feel to the country, having recovered from threatened invasion followed by the Six-Day War. Our Israeli tour guide stopped at many locations to say prayers for his fellow soldiers who had perished in that war. We saw the first moon landing on a television whilst there – albeit in black and white. The student hostels we stayed in had no air conditioning which made coping with temperatures of up to 45 °C pretty uncomfortable.

On our return I happened to tell Tricia about the 'Elephant Man' – a very cruel and pejorative description of the shy yet artistically talented Joseph Merrick, born in 1862, who developed extensive skull, bony and skin deformities by the age of five. His skeleton and the cowl he wore in public were on display in the medical college pathology museum. I introduced Tricia to his skeletal remains and grotesque visor. I suppose we were no better than the general public in Joseph Merrick's day, who came to view him as a mawkish curiosity in circuses. He was later managed as a touring exhibit. There is some evidence suggesting that Merrick suffered from the Proteus syndrome and had the following features compatible with this diagnosis: macrocephaly (large head), bony protuberances of the skull and long bones; thickened skin and subcutaneous tissues, particularly of the hands and feet, lipomas (benign fatty tumour) and other unspecified subcutaneous growths.

After an extremely difficult life he eventually found some degree of peace in his final years, being looked after and cared for in an attic at the London Hospital by the eminent London Hospital surgeon Sir Frederick Treves, who himself went on to save the life of Edward VII. The new monarch who had just ascended to the throne in 1901 after the death of Queen Victoria his mother, needed an appendicectomy just before his coronation, at a time when this operation was still very new and untried. The King was soon up and about and able to be crowned thanks to Treves.

John Merrick

Public domain

Back to John Merrick. In his last few years he was able to undertake impressive artwork and was feted rather than exhibited, being visited by the great and good of the age. I believe his skeleton is now somewhere in Queen Mary College, which is now part of the University of London. It is used for teaching purposes only.

And so to finals, which took place in the early summer of 1971. I have copies of the written papers in Medicine, Surgery and Therapeutics. They were essay style, usually with a choice of answering four out of five questions or thereabouts. I met up with Dr Henson again who was partnered by an external examiner. The patient I examined had neurological symptoms and signs, and I deduced he had hereditary cerebellar ataxia. I martialled my thoughts in what I hoped was some sort of logical order to present his neurological condition, which was a genetically caused tremor. He incidentally had chronic bronchitis. The examiners were only really interested in his chest problems but just before the end they asked me what I made of his neurological symptoms, which I proudly detailed. I cannot remember much at all of the surgery exam. One of the gynaecology examiners spent an inordinate amount of the oral exam time commenting on my stammer, which of course was totally inappropriate but quite usual for the times. He implied that doctors should be robust in body and mind, and that the infirm should be blown away, like so much chaff in the winds that sort the able-bodied from the enfeebled.

Results day arrived. We were given them informally over a pint of beer in the students union by Mike Floyer who was both professor of medicine and a keen educationalist and went on to be the Dean of the Medical College. I was told I had only just missed an honours examination in medicine by a mark. And so here I was aged 23 a doctor. My next hurdle was to do pre-registration house jobs in surgery and medicine – six months each.

Chapter 6

LIFE AS A JUNIOR DOCTOR

An aside now about my personal life. Tricia and I were married in November 1970 at Dunstan Road Synagogue, the same place as I had my bar mitzvah. After the wedding there was a large sit down dinner in the Mount Royal Hotel in London's West End. Tricia and I spent the first night of married life in the Royal Garden Hotel in Kensington and then flew off in a BEA Trident to honeymoon to Gibraltar. The weather there was generally poor. We took a short flight to Tangiers, accompanied by two sheep in the cabin. In Tangiers we enjoyed the Moroccan market including seeing a number of snake charmers. Back home in London a week later, we started married life renting a flat in Hornsey. It was an upstairs apartment in a terraced Victorian house. Hot water for a bath came from a 1930s Ascot gas heater, which in retrospect was a health hazard as well as delivering water at a trickle. We stayed there about 18 months. Tricia always recalls that she was very happy there. We had been offered a small house in Turner Street next to the London Hospital for £500 but felt we could not quite afford to buy it. It is now worth £1.6 million!

We both were unsure that we would ever be able to have children, so used no contraception. Tricia's concern was irrational. Mine had a basis in fact. My concern about fertility was based on the fact I have, and still have, a left-sided varicocele. These are a collection of varicose veins in the scrotum, which Bailey and Love, my surgical textbook informed me, was associated with infertility.

The Greek physician Celsus in the first century AD gave an accurate description of varicoceles. The scrotum feels like a 'bag of worms'. It is generally accepted that 20 per cent of men with varicoceles have fertility issues and that 40 per cent of men with fertility issues have a varicocele. 90 per cent of varicoceles are left sided. This has been assumed to be because the left spermatic vein is 'nutcrackered' between the aorta and the upper mesenteric artery. The exact mechanism of how varicoceles cause low sperm counts is unclear – increased temperature, release of adrenaline like substances near the testis and so called 'oxidative stress' are but some hypotheses. Removal of the varicose veins has been thought to improve male fertility. However one study showed that the fertility rate of couples a year after surgical intervention was no different from couples where the man had a varicocele and only received counselling.

Varicoceles are by no means the only cause of low sperm count. A literature review in 2017 found that sperm counts in men not known to be infertile had fallen by up to an average of 60 per cent in North America, Europe and Australasia between 1973 and 2011, most likely due to 'environmental factors'.

Kantartzi, PD, Goulis, ChD, Goulis, GD and Papadimas, I. Male infertility and varicocele: myths and reality. Hippokratia. 2007 Jul;11(3):99–104.

Levine, H, Jørgensen, N, Martino-Andrade, A, Mendiola, J, Weksler-Derri, D, Mindlis, I, Pinotti, R and Swan, SH. Temporal trends in sperm count: a systematic review and meta-regression analysis. Human Reproduction Update. 2017 Nov 1;23(6):646–59.

The sperm analysis I produced for assessment showed a low count. However, our fears were allayed when Tricia became pregnant after a month! We went on to have seven more children. So if you have a varicocele and a somewhat low sperm count, don't be disheartened!

And so to the preregistration house jobs. Each one lasted six months. Nowadays the newly qualified doctor rotates through more than two specialties in their so-called Foundation years, which often include general practice, each for a period of four months.

My first house job was in paediatrics at Mile End hospital. I started the job in the high summer of 1971, when the pollen count peaked. I was sneezing so much I tended to have nosebleeds. While frantically attempting to stop the blood pouring out of my nose, I was called to see an eight-year-old girl who had acute leukaemia. She was a beautiful child with long blonde hair, blue eyes and a pleasant demeanour. I had attended her on many mornings when she also had quite severe nosebleeds because her blood platelets were almost non existent. So there I was packing her nose with one hand while holding my nose with the other. She was delighted that we both had the same problem and I guess it was a bonding exercise. The prognosis for acute childhood leukaemias in 1971 was poor. I left Mile End while she was still alive and continuing with her early morning nosebleeds. While at Mile End I was the sole 'paediatrician' on call at night and weekends. The reality of the situation was that I was a just qualified, very junior doctor, who knew very little about children's health. In spite of this I was often called by A and E for clinical opinions, which of course was laughable – it was more like the blind leading the blind. I had a big textbook in the room I slept in and desperately turned the pages of that tome to attempt to find some sort of answer before going to A and E. Of course those were the days before Dr Google was available! There were more senior paediatricians I could call on but they encouraged me to try to handle cases myself. A memorable problem was being asked about a toddler who had ingested all his mother's oral contraceptives. I could not contact the paediatric registrar on call and could find nothing in the textbook on the topic, so I reassured the A and E doctor that all would be well. If the child had been a girl I guess she might have had a withdrawal bleed a few days after taking the pills. The little boy may have grown transient breasts. I never knew the outcome but surmise that I gave the correct advice. Tricia stayed with me over long weekends on call as she was heavily pregnant. After three months at Mile End I transferred to the London. The inpatient paediatric unit was on the second floor in the middle section of the hospital. The unit had wooden doors each end with the handle high up on the doorframe. There were originally two consultants. Dr Richard Dobbs was on the point of retirement and was a pleasant, friendly and very experienced general paediatrician who wore a monocle. The evening before one of his ward rounds I admitted a ten-year-old boy who was in severe pain in multiple joints. Although there appeared to be no joint swelling, I felt I ought to bring him in for observation. Dr Dobbs examined him the next day and felt pretty sure he was suffering from 'growing pains'. I felt I had rather overreacted, but Dr Dobbs was very kind about my misdiagnosis saying it was an understandable error. Dr Jackson, an eminent paediatrician, was in his 50s, a tall rather irascible man. On one occasion he shook a child who would not keep still while he examined him. Ward rounds with both consultants could last from morning till early afternoon. At lunchtime the whole team repaired to the pub, had a light meal as well as one or two pints of beer each. How times have changed!

There was better senior cover while I was on call at the London. In spite of this I was often summoned to the maternity unit when it was anticipated a newborn might be in trouble at and after the birth. This would entail placing the neonate on a heated resuscitation trolley head down and giving oxygen by facemask. If the baby's response to these basic manoeuvres was suboptimal, I would proceed to intubation, that is passing a plastic tube directly into the neonate's airway and oxygenating them till a full response took place. And this is what happened when my son Oliver was born. I was the paediatrician on call, and he needed intubation! A rather unusual introduction to my firstborn!

In the last few months of the paediatric job I had to think about what job to do next. I had enjoyed surgery at St Margaret's Hospital in Epping. I phoned one of the surgeons, Mr Pelham Warren who remembered me as a student. He told me to come to the operating theatres at about teatime on a specific day. He came out in his green theatre scrubs, said hello with a smile and asked me some questions with his usual nasal intonation. He was a tall, lean man of few words. After 10 minutes the job was mine! No formalities, CV, group presentations, psychometric testing or solo presentation or indeed competition with other candidates – just accepting me after a short chat!

By the time I started the surgical job, my little family had moved to a modern flat in Wanstead. I drove my 800 cc, two door, white Austin Mini (the original type, with a cord rather than handle to open the passenger door!), past the bronze statue of Winston Churchill in his old constituency of Woodford Green onto Epping, an easy journey, going the opposite way to rush hour traffic. The hospital buildings at the time I worked there had originally been part of the Epping workhouse. They were converted into a hospital in 1938 and became part of the NHS in 1948. The hospital water tower was hit by a V2 rocket in 1945 killing seven men. Each ward consisted of a large metal Nissen building, the inside being in Nightingale style, that is, a large open plan area with two rows of beds facing each other. The sexes were segregated in separate ward buildings which were named after common trees. Each ward was controlled by a ward sister. Cherry ward's nursing sister made it quite clear to me from the outset that her ward nursing routines were not to be disturbed in any way by my presence. The patients were her domain. I was in large part treating her patients under sufferance. Of course the real treatment, as she saw it, was undertaken by the consultants, with her in charge of their day to day recovery.

Almost all of this old hospital has since been demolished and a new purpose-built structure erected. I was not particularly interested in surgery. The surgical registrar, who had passed his FRCS examination and was of middle rank, was a Munsah Salah, who at some stage had Palestinian connections. He was generally pleasant enough to me but became very heated when talking about Israeli Palestinian politics. I tried not to engage him in such topics, particularly while he operated, as he had a tendency to point his scalpel at me in anger as he emphasised his point of view. The nursing sister in charge of Rowan surgical ward had a dense middle European accent and seemed very obsessed with bowel evacuation – not surprisingly as a lot of the surgery was abdominal. Bowel movements were considered almost holy, and included using Celevac – I can still hear her intoning that name after nearly 50 years! I was told by all the staff not to engage with the patient in the first bed. He looked pale, sallow and was cold to the touch. Apparently he had developed a septicaemia and was now waiting to die. He was ignored by the surgical team. I tried to engage him in conversation. However he was too sick to answer me or perhaps did not want to. He died three weeks after I joined the team.

On the day prior to each surgical operating session, I assessed the patients' medical health and anaesthetic risk. Nowadays this is done by the whole anaesthetic team including a consultant, but then the onus was devolved to a very junior doctor. After screening all the patients by taking a history and examination, I discussed them over the phone with the consultant anaesthetist, who told me what to prescribe as a pre-med – usually included omnopon which was a mixture of opiates. In the operating theatre, when called to assist, my usual task was to keep the abdomen open with a retractor, while Mr Warren delved into the guts. Surgical removal of stomach ulcers was not uncommon. As I discussed previously, with the advent of powerful medication to stop stomach acid production (e.g. ranitidine or later on omeprazole) these operations were just not necessary and became largely redundant. Mastectomies and large bowel cancer were other typical operations. On one occasion my glasses fell into the open abdominal cavity. Mr Warren just picked them up with forceps, facetiously asked me if they were mine, and then told me the name of a good local optician!

I was encouraged to undertake an appendicectomy on my own but I called for help as I could not flip out the offending organ. I did do minor outpatient surgeries such as removing benign cysts or ingrowing toenails. Mr Warren was an excellent surgeon but a man of few words. If a patient admitted under him did not require surgical intervention he would nod at them, grunt but say nothing at all to them. A woman with gonococcal inflammation of her fallopian tubes causing lower abdominal pain was thus passed by at the ward round. My task was to ask for advice on her treatment from Dr Titus Oates, the venereologist who worked in the locality.

Towards the end of the six months with the surgeons I approached the adult physicians to see if I could do six months with them, in what was general medicine. This was as a post registration house officer. I was interviewed, more formally this time, and got the job. At that time the management of common serious medical (rather than surgical) conditions such as myocardial infarcts (heart attacks) was much less advanced than it is nowadays. Today the blocked coronary vessel is dilated at angioplasty at the time of presentation. Back then, much of the treatment was to control the symptoms such as pain with morphine, to boost the heart if it was failing with digoxin, to control rhythm problems with propranolol and to rid the body of excessive fluid with diuretics such as frusemide. A pleasant married middle aged man who suffered a heart attack had to be resuscitated a number of times after his heart developed a rhythm problem. He worried that the love letters from his extramarital affair, which he had brought with him to the hospital, would be read by his wife in the event of his death. He insisted that I should remove and destroy them in the event of his death. He fortunately made a good recovery and was discharged home. Another patient was a young man who was performing at a local fair and drove his motorbike through a tunnel of fire – except he did not make it. By the time I saw him in A and E he was just a charcoaled mass.

I worked for two consultants in Epping in general medicine – Dr Medley and Dr Harry Kopelman. Dr Kopelman, a short man with a round face, combed back black hair and a definite air of authority, had written a paper titled 'The Epping Jaundice' in 1966, describing 84 local residents plus a medical student who had developed abdominal pain and jaundice. After some detective work it was found that a chemical diaminodiphenylmethane had leaked from a container over a bag of wholemeal flour while the latter was being transported to the bakery. All the patients involved had eaten the subsequently baked wholemeal bread.

In those days junior doctors did one in two evenings and alternate weekends on call. The weekend was long – lasting from Friday morning to Monday evening. There were fewer complex medical interventions in those days, so patients could be clerked in, treated and observed or discharged relatively quickly. Whilst on call all meals were free and there was a great atmosphere of camaraderie amongst all the various medical, surgical and gynaecological teams. Whilst on call one could leave the hospital grounds so long as you were not too far way. The bleeps we carried received messages from about 2 miles away. On one occasion there was little activity whilst I was on call over a weekend. I therefore quickly visited a house party in Epping Town. Just as I arrived my bleep went off. I was needed urgently for an emergency. I responded rapidly, driving over the speed limit to the hospital. En route I was stopped by a police car. Having told them the facts of the situation they drove ahead of me with lights flashing, escorting me through the night traffic to the hospital.

A short distance from the hospital lies North Weald airfield. In the early 1970s airshows were still held there. As a doctor working in the local hospital I was given a grandstand view of the events there (again while on call). The airfield played a major role in the Battle of Britain, being home to some of the RAF squadrons made up of Czechs, Poles and Norwegian fighter pilots.

Chapter 7

WHITECHAPEL CLINIC

Towards the end of 1972 I had to think about my next job and indeed what medical career I wanted to pursue. General medicine seemed attractive. I therefore applied to be senior house officer in general medicine (in the USA general medicine is termed Internal medicine) at my alma mater. The job interview was a stammering debacle. Four stern faces glowering at me. One of them asked me the usual question of why I wanted the job. I came prepared for that question. There was a logical answer. I wanted to be taught by great physicians in a hospital that has a national reputation for excellence. But the words wouldn't form. All my effort went into verbal articulation. I lost my train of thought. I floundered in a sea of anger, exasperated at letting success slip through my fingers, just because I couldn't enunciate the words. My fear and annoyance further clouded my train of thoughts. I looked at the interviewers. They had already decided they didn't want me. It was a waste of time carrying on with the interview. Why would they want to burden themselves with a junior doctor who takes three times as long as an average person to deliberate and convey information? Easier for them to say no. The interview ended. I turned away, rose and walked out of the door. I am so ashamed of myself. I noticed my heart was beating fast and my hands were clammy. In years to come I would wonder what my life would have been like if I had interviewed better? Come to think of it, my daydreaming continued, what if I never had a stammer? I told myself that I may have been a slicker, faster thinking doctor, maybe a wonderful orator, but on the downside a haughtier, less empathic doctor. OK, so my stammer made me a kinder person. More understanding. How much would it be worth to pay the devil to rid me of it? Shades of 'Dorian Gray'? The fantasy faded and I remembered that I did not get that job. My stammer had once again badly let me down. All the candidates apart from the successful one were thanked for attending and dismissed. But as I walked away I heard, 'A moment please Dr Goldmeier'. Dr Philip Rodin, a consultant venereologist on the interview panel asked if I wanted a job as a junior in his department – called the Whitechapel clinic. Venereology literally means the science of the act of love, which clearly explained the speciality. However, the term became pejorative in the 1960s and 1970s, carrying with it implications of having a self-inflicted disease that was treated well away from the rest of the hospital, usually in dark basements. In spite of any misgivings, and because I had a growing family to support, I accepted the offer – little knowing that a long career in sexual health and sexual medicine would follow on from that one job. I initially began working in the unit as a senior houseman, and later became a 'supernumerary' registrar followed by a post doing research work for an MD degree.

In order to eventually become a consultant in a medical (rather than surgical or gynaecological) speciality or general practice, one had and still has to pass the MRCP (Member of the Royal College of Physicians) examination. The MRCP is a test of advanced general medicine and had both written and clinical parts. Part one consisted of multiple choice questions and preparation for it entailed reading textbooks and memorising and understanding the information therein. A number of junior doctors at the London were taking the exam contemporaneously. We therefore tested each other in the evenings after clinic or at lunchtimes. I passed first time in 1973. Part two largely involved clinical examinations. Passing that exam involved the appropriate knowledge, obtaining an accurate history from the patient and examining her/him to obtain clinical signs. All this together would lead to the correct diagnosis or at least a differential diagnosis, that is, the various reasoned possibilities of the patient's illness. The case had then to be presented using optimal medical language that was logical, terse and to the point. The latter could not be found in books and was not taught well at hospital ward rounds. That being the case, I went onto a course specifically to prepare for the clinical exam. The best one by far at that time was given by Dr Maurice Pappworth. It took place in Seymour Hall in London's West End. He lectured to large groups as well as taking participants to 'geriatric hospitals', where the patients had many clinical signs. His clinical examinations were undertaken 'one to one', and were almost as nerve-racking as the actual MRCP exam itself. However it did teach me how to elicit clinical signs and present the findings with minimal verbiage, yet with the clearest meaning. An example of Pappworth's brief, yet informative medical terminology, is a description of a third cranial nerve palsy. What the clinician observes is that the patient is unable to open their eyelid. Beneath the lid, when withdrawn, is an eyeball that is fixed downwards and outwards. The pupil is large and does not respond to a light shone at it. All rather verbose! The Pappworth description is 'internal and external ophthalmoplegia' – meaning there is partial paralysis within the eyeball, as well as of the external muscles of the eye. MRCP examiners could spot a 'Pappworth candidate' by their accurate and pithy replies. Pappworth also gave social instruction to foreign candidates, such as not to make snorting or coughing noises during the MRCP examination. This was good advice for those born and brought up in foreign climes, whose cultural mannerisms did not fit in with the idea of an English physician. Nowadays, in our woke era, this advice might be construed as racist. But Pappworth was only trying to help. I know I was grateful to him.

Because of him I passed part two first time round.

Maurice Pappworth was a practising orthodox Jew. He came up against anti-Semitism early on in his career, so that he was unable to obtain a consultant appointment either before World War 2, during which he was promoted to Lieutenant Colonel, or after the war. He was told 'A Jew could never be a gentleman'. He openly criticised the medical establishment, in particular in relation to unethical practices of the day regarding medical research. His published articles and appearances in the media of the 1960s and 1970s were instrumental in the establishment of ethics committees, which oversaw research studies on humans. His opinion on unethical medical research found expression in his book Human Guinea Pigs. *He was a bête noire as far as the Royal College of Physicians was concerned. As the older cohort of establishment physicians died out, new blood at the College eventually feted him by virtue of his hugely successful MRCP courses and ethical ideas, and he was belatedly made a Fellow in the 1990s.*

The Whitechapel clinic, the eponymous name of the venereology clinic, was housed, as I explained above, in a dungeon of a place beneath the living accommodation for staff on call for the London hospital. As if to dissociate itself even further from what was construed as unseemly medicine, the department also lay to one side of the main part of hospital, maybe so as not to disturb the sensibilities of those practising 'real' medicine. The senior consultant was Dr Eric Dunlop, who spent most of his day in his office from where he was 'consulted' by the requests of junior doctors or the more senior nurses. I believe he had a large private practice.

He was born in Argentina in 1923 and qualified at St Bartholomew's hospital in the early 1950s and became a consultant venereologist to the London Hospital in 1954 and to Moorfields Eye Hospital in 1964. He was one of the pioneers in working out that chlamydia trachomatis (CT) was a genital pathogen. The man's academic writing style was dry but very accurate. He was referred to as 'Eric' behind his back. His interactions with staff of all levels, from the most junior nurse to consultant colleagues, could be described as medically feudalistic, where he as the master dictated orders and gave opinions on all matters. So long as one had not committed a misdemeanour, he would speak with a part dominant, part polite smile. On other occasions this rather tyrannical nature could transform into a more gentle soul, who was polite, gracious and kind. I became his research registrar 18 months after joining his department while studying for an MD (medical PhD). On the first day of that job he told me I would be giving a lecture to a postgraduate course on mycoplasmas, as he had to be elsewhere. Mycoplasmas are a group of bacteria, some of which cause genital disease and others merely act as genital passengers without causing disease. Many questions hang over the pathogenicity of the mycoplasmas to this day. There was even more uncertainty in the early 1970s. He should have given this lecture. I was given two days to prepare for the talk. No time to make 35 mm slides. The only audio-visual aid I had was chalk and a blackboard. The talk was not a great success – my stammer has always been worse when I am unsure of the facts I am speaking of. I did ask him for help with the talk. However he did not make himself available. Eric was an insulin-dependent diabetic. At times this was poorly controlled, so that I or other members of staff had to rescue him from talking gibberish to medical students just before lunchtimes when his glucose levels plummeted and he became hypoglycaemic. He undertook research work with Prof Barry Jones of Moorfields Eye Hospital and the Institute of Ophthalmology on CT. I helped him run his clinic at Moorfields, where patients presenting with chlamydial conjunctivitis as well as the occasional patients with syphilis or gonorrhoea had presented initially with ocular problems. They were then transferred from the main eye hospital to have their concomitant genital infections diagnosed and treated. That clinic ran with the help of Russell, a tall, slim, middle-aged charge nurse. He always seemed to be content with life, speaking to patients and staff alike with a BBC English accent in a mellifluous voice, punctuated by a loud fruity laugh. Russell would intermittently comb his hand through his long greying black hair as it obstructed his efforts to view the genitals and take chlamydial tests from the penis. His major love was the stock market. If a patient had knowledge of stocks or shares, his eyes would light up and he would interrogate them while taking their genital tests. The patients' answers were generally garbled by virtue of their contorted faces and groaning pains as he took the samples! He was a chain smoker whose cigarettes never left his mouth, and often collected long centimetres of ash that would spill onto the patient and examination couch. CT samples from the penile urethra were taken by means of a curette, which was a long thin metallic scraper, plus a swab, which was a wire with cotton wool at one end. Prior to their insertion, which was to the full length of the penis, liquid anaesthetic (lidocaine) was squirted into the penile urethra and the penis tip squeezed so that the anaesthetic could take effect. Most patients not surprisingly found scraping their urethras pretty painful. Nowadays all that is needed to detect chlamydial DNA is a first catch urine.

CASE – CHLAMYDIA TRACHOMATIS

The year was 1976. The woman was deep in conversation with Russell my charge nurse colleague as I approached. He had noticed her reading the Financial Times *in the waiting room and was now in my clinic room, pleasantly but persistently quizzing her on details of the latest stock market movements. If he had been more interested in her appearance rather than her knowledge of securities, he might have noticed, as I did, that she was a long boned, muscular woman with full lips, a delicate snub nose and short, layered black hair. As I sat down and introduced myself she told me in a gravelly voice that her name was Sheena. She was wearing shiny lilac ankle length tights topped by a florally decorated sleeveless T-shirt that descended to mid-thigh level and a pink head sweat band. She had been referred to the euphemistically named 'Diagnostic Clinic' at Moorfields Eye Hospital, which was in fact a sexually transmitted diseases clinic that had changed its title so as not to offend the genteel and to save sensibilities. Sheena was a 44-year-old fitness instructor who presented with a left, red, weeping eye to Moorfields A and E department. She was found to have CT causing the conjunctivitis and was treated with appropriate eye ointment and subsequently seen in the Diagnostic Clinic. Sheena had two recent sexual partners – both men. A year before she had undergone a total hysterectomy because of fibroids causing heavy bleeding. A swab was taken from her upper vagina. It confirmed a CT infection and she was treated with a week of antibiotics – doxycycline. She could only inform one of the partners and urged him to get tested and treated. She could not trace the other man.*

Chlamydia trachomatis (CT) is a bacterium, although it was originally thought to be a virus. The etymology of the word 'chlamydia' derives from the Latin for a cloak, as the nucleus of the infected human cell appears to be covered by what was thought to be a dark looking, large parasite. There are a number of different subtypes or serovars of CT, some causing inflammation of the urethra and testes in men and the cervix and fallopian tubes in women, others a non-sexually transmitted eye infection termed trachoma which causes corneal scarring and is a leading cause of infective blindness in Africa, transmitted by fly vectors. Yet further, rarer, serovars cause severe rectal inflammation and enlarged matted lymph glands, termed lymphogranuloma venereum (LGV), which is sexually transmitted. CT in the guise of genital infection is the commonest sexually transmitted disease in the UK. It may however cause no symptoms at all.

Before the 1980s the received wisdom, as stated above, was that the cervix had to be present in order for the woman to become infected, although CT had also been found in the urethra and rectum. However in 1985 I and colleagues described four women who had undergone hysterectomies and yet were infected with CT in their vaginas. Detection of CT was confirmed by staining vaginal exudate with a monoclonal antibody against the organism. The most likely explanation is that the primary infection was in the urethra and that the vaginal was secondarily contaminated by urine seeping into the vagina. However it is possible that there was direct infection of the vaginal cells. Nowadays, using much more sensitive DNA technology, the site of choice for CT testing in women is the vagina. CT in that location may derive from the cervix, urethra or rectum, or as we demonstrated nearly 40 years before, the vagina itself.

What was and remains certain is that the conjunctiva can be directly infected by CT. In Sheena's case it may have been by digital contamination directly from a sexual partner or rather more likely from digital contamination from her own genitals.

One of the co-authors of my research work was Professor David Taylor Robinson who originally worked in the Common Cold unit in Porton and later at Imperial College. In 1980 his lab isolated Mycoplasma genitalium, a hitherto unknown bacterial genital pathogen. It took all of 80 days for a significant growth to appear, whereas most bacteria grow in a few days. David was a tenacious researcher with a puckish sense of humour. He was asked to give a talk at a prestigious meeting titled 'Does chlamydia cause rheumatoid arthritis?'. He was silent for well-nigh a minute, then spoke one word – 'No' and left the podium. The audience were appropriately shocked. A minute later he returned to the lectern to give a 40-minute in depth talk detailing the evidence for his single word answer.

Schachter, J, Chernesky, MA and Willis, DE et al. Vaginal swabs are the specimens of choice when screening for Chlamydia trachomatis and Neisseria gonorrhoeae: results from a multicenter evaluation of the APTIMA assays for both infections. Sexually Transmitted Diseases. 2005 Dec;32(12):725–8.

Tully, JG, Taylor-Robinson, D, Cole, RM and Rose, DL. A newly discovered mycoplasma in the human urogenital tract. The Lancet. 1981 Jun 13;1(8233):1288–91.

Barton, SE, Thomas, BJ, Taylor-Robinson, D and Goldmeier D. Detection of Chlamydia trachomatis in the vaginal vault of women who have had hysterectomies. British Medical Journal (Clinical Research Edition). 1985 Jul 27;291(6490):250.

At the Whitechapel clinic the junior consultant Dr Philip Rodin, who was editor of a national venereology journal, complained bitterly to me off and on, about how badly he was treated by Eric. The day to day running of the unit was undertaken by the junior doctors and supervised by the senior charge nurse Mr 'Goldy' Goldsmith, who taught me a lot about the fine art of smoothly handling the sensibilities of both colleagues and patients. Goldie was a soft spoken, small, somewhat rotund man in his fifties, with jet shiny black hair, and a generous smile, who bustled quietly around us, ensuring as far as possible, the progress of each day's clinic went without a hitch. He was a single man, who had been called up to National Service in the 1950s and had a rather overexaggerated moral compass. When he saw me chatting for what he perceived to be too long with an attractive nurse he would interrupt the conversation, telling me later on it was not appropriate for a married man to so spend precious clinical time talking to someone of the opposite sex. Dr Tom Stuttaford worked sessions in the clinic with me whilst still a Conservative MP for the Isle of Ely, having won the seat back from the Liberal TV broadcaster Clement Freud in 1970. After losing the seat in 1975 he took over my research job as well as being a private GP. He went on to be the doctor writing a regular medical column for *The Times*.

Until my time at the Whitechapel clinic, my medical examinations had generally not included assessment of the male and female genitalia. They are of course done with gloved hands. Goldy, rather than any doctor, explained how one should conduct and go about a male examination. Some of the more experienced junior doctors also gave of their time. Seemingly it was beneath the dignity of the consultants to so do. At medical school we were taught the sequence that one should use to clinically examine any part of the body – inspection, palpation, percussion and finally auscultation. The latter two were rarely necessary in genital examination. 'Keep the man standing throughout the examination'. That said, the man's dignity can be lost with his trousers and underpants as they come to rest just above his shoes. One of my consultant colleagues in later years would never let men sit down, as he said it would retard the general clinic flow by wasting precious time getting the patient to lie down and then stand up again. Inspection may reveal obvious abnormalities such as an ulcer on the penile shaft, a dripping urethral discharge or an enlarged testis. Noticing if the patient is in pain goes without saying. On to palpation. The foreskin needs to be withdrawn. An ulcer or cancer can lie beneath. If no discharge is spontaneously seen then the urethra is milked from above to below to view any material lurking unseen within the urethra. The testes are then felt by gently cupping each in the palm of the examining hand and proceeding to gently feel for enlargement and any hardness or lumps. Lastly the groin is assessed for any enlarged glands or hernias. If the abdomen needs examining the patient is asked to lie down. All easy enough you may say. Some foreskins are tight and wont retract. Penises can misbehave even when their owner's best intentions are to co-operate with the examiner. An unwanted erection. This can be just as alarming for the inexperienced doctor as for the patient. Reassurance that this is normal is the order of the day. However, some men may 'try it on', in particular saying that they do not want a man to examine them. In my experience gay men are no more likely to suffer an involuntary erection when examined by a man than a straight patient.

I had met the female genitalia at examination before, while doing obstetrics and gynaecology as a student. Dr Patsy Pereira, an almost skeletally thin Anglo-Indian, the senior registrar at the Whitechapel clinic, was very pleasant and generous with her time and patience, and taught me a great deal about female examination. The necessity for careful inspection and palpation of the woman inevitably means that her dignity is compromised. Women must take off all lower clothes and panties. The external genitalia are most accurately assessed with the woman's feet in stirrups or the back of her thighs on rests. Legs have to be opened (abducted). Placing a paper towel over the lower abdomen affords her some degree of propriety. The nurse should be as much present for the patient as being a chaperone and helper for the examiner. The outer lips (labia majora) and then the inner lips are carefully inspected, and then a speculum is inserted into the vagina and opened out, so as to inspect the vaginal walls and cervix. Specula used to be made of metal and were sterilised after each use. On winter days they needed warming up prior to insertion. Nowadays single-use plastic ones are the norm. Contracted pelvic floor muscles can make passing the speculum either painful or impossible. Sparing use of lubrication such as KY Jelly helps to make insertion easier for the woman. Too much lubrication inhibits reading the microscope slides as the bacteria and fungi are made less easily visible. The vaginal examination completed, two lubricated fingers are inserted into the vagina, lifting the uterus at the cervix and with the other hand on the abdominal wall balloting – meaning feeling – the uterus and ovaries between the two hands. Before withdrawing the fingers the cervix is moved from side to side. If there is inflammation of the uterus or fallopian tubes this manoeuvre is painful, even with the smallest amount of pressure being exerted on the cervix.

FOUR CONDITIONS SEEN IN EARLY 1970s

VAGINISMUS

I met Christine in 1973 at the Whitechapel clinic. She was 19 and had attended complaining of a copious yellow vaginal discharge. Because she was also highly anxious about being examined she had brought along a female friend for support. I was working with Wendy Roland. Wendy was the senior female nurse in the department at that time. She was an attractive, no nonsense, pleasant, compassionate young sister whose rugby-playing boyfriend was her pride and joy. I attempted to insert a pre-warmed, lubricated speculum but Christine's legs closed with fear and she pushed me away, almost screaming that she couldn't cope. I suggested that we give her some diazepam (a tranquiliser) intravenously to relax her. Nowadays, one would have to obtain written consent for so doing, and the procedure would be considered an anaesthetic. Nevertheless with youthful optimism and gung ho on my part I proceeded. Diazepam given, I inserted the speculum with ease. Christine voiced a subdued moan as I did so, which I dismissed. However, she then began thrusting rhythmically against the inserted instrument. I attempted to reassure all in the room that I was in control of the situation 'examination nearly over'. She then announced, 'It feels great George, do it harder', to which her friend replied, 'It's not George it's the doctor' 'I don't care, just do it harder'. At this point Sister Rowland's face turned a deep shade of crimson as she eyed the ceiling. Christine's friend looked at the floor. I looked at Christine's cervix. I obtained the specimens I needed. She turned out to have gonorrhoea. When she came round she expressed gratitude that the examination had gone so well but seemed not to recall her words or manoeuvres. Whenever I met Wendy Roland after that event, she flushed with embarrassment! I think this was the first case of vaginismus I saw. More of that later.

BODILY ODOURS

Many women come to a clinic complaining of a smelly vaginal discharge. Surprisingly CT and gonorrhoea discharges are not particularly offensive. A very common cause is bacterial vaginosis (BV) which occurs when there is a deficiency of the healthy vaginal flora of lactobacilli and an over growth of multiple organisms that thrive without oxygen such as Gardnerella and Mobiluncus. However, in the early 1970s BV had yet to be discovered. When no obvious cause of a smelly vaginal discharge is found, it is worth looking behind the cervix, where a forgotten tampon may be lurking in a greyish brown ball that exudes a powerfully putrid odour. Once seen, the examiner needs to alert their colleague. Forceps are used to capture and remove said object – hopefully all in one piece! It is immediately placed in a paper bag to minimise the spread of the offensive odour. However even those few seconds in the air are enough to render the room unusable for some time.

A much rarer cause of offensive body odour is trimethylaminuria, where there is an enzyme defect, so that the smelly chemical trimethylamine, which is produced in the large bowel by the degradation of foods with high amine content, accumulates in the body. It should be broken down physiologically by a specific enzyme to an odourless compound. Trimethylamine is excreted in sweat and vaginal transudate. There is no specific treatment for the condition. However cutting certain foods from the diet can be helpful as well as laxatives so that food and faeces does not linger in the large bowel.

https://www.mayoclinic.org/diseases-conditions/bacterial-vaginosis/symptoms-causes/syc-20352279

Roddy, D, McCarthy, P, Nerney, D, Mulligan-Rabbitt, J, Smith, E and Treacy, EP. Impact of trimethylaminuria on daily psychosocial functioning. JIMD Reports 2020 Oct 6;57(1):67–75.

Bagnall, P and Rizzolo, D. Bacterial vaginosis: A practical review. JAAPA. 2017 Dec;30(12):15–21.

GENITAL HERPES

When I saw my first patient with genital ulcers I had to once again enlist Goldy's help to give me a brief summary of genital herpes before I read up on it later on. The male patient in question identified himself as 'homosexual'. The term was somewhat antiquated even in the early 1970s. In the 1950s up to 1,000 men were convicted and imprisoned in England and Wales because they had committed 'homosexual acts'. The 1957 Wolfenden Report recommended that homosexual acts between men in private was not a matter for the law. This edict was enshrined in the law only ten years later. So by the time I was started practicing venereology, homosexual acts were no longer illegal. The term 'gay' entered the general vernacular some years later. Recent UK surveys suggest that about 3 per cent of the male population identify as gay. However up to 10 per cent of men may have sex with another man, but do not identify as gay. Now, until this point in my life I had never knowingly met anyone who identified as gay. The prevalence figures as stated above would have meant at least a handful of boys in my school year were gay. But they were well hidden for good reason. So here was this young man, in obvious pain, with multiple ulcers on his penis and anal area. They were exquisitely tender to touch. Goldy informed me that even though this was 99 per cent likely to be genital herpes, I should take tests for syphilis as well. Syphilitic ulcers are harder and tend to be painless. I took two cotton bud swabs to send off to the lab to confirm genital herpes, one from the anal area and one from his penis. Each time the swab touched the ulcer, he screamed and almost jumped off the bed. We gave him some ibuprofen tablets to dampen the pain and kept him under observation for a few hours, during which time he was unable to pass urine. An enlarged bladder was later confirmed examining his abdomen. By the end of the day he had not passed urine, so he was admitted to the wards, where a urinary catheter was inserted and kept in place for a few days. He was also unable to open his bowel for ten days. The patient had a stormy passage over the next two weeks, but eventually all the ulcers healed. The early 1970s was prior to the specific anti-herpes drug acyclovir becoming available. I wrote my first ever research paper in 1973- describing the above patient and a similar case seen four months later. Both men suffered from severe anogenital herpes simplex and who went on to develop intestinal obstruction – that is they were unable to open their bowels for a number of days. The bladder and bowel dysfunction may have been the result of the virus causing painful ulcers at the anal margin resulting in reflex spasm (contraction) of the anal muscles thus preventing defaecation. However the virus is 'neurotropic' – that is, taken up by local nerves. That being so, the nerves that supply the pelvic floor muscles may have been affected, thereby not allowing messages to those muscles to relax to empty bowel and bladder.

Goldmeier, D, Bateman, JR and Rodin, P. Urinary retention and intestinal obstruction associated with anorectal Herpes simplex virus infection. British Medical Journal 1975 May 24;2(5968):425.

SYPHILIS

There had been a large upsurge in syphilis in the UK during both World Wars. It often presents in its early stages with a genital ulcer or a rash, when treatment is given to prevent it going onto the tertiary or final destructive phase. However it can be clinically silent and only detected when damage has already taken place. At the Whitechapel clinic we saw patients who had tertiary or end stage syphilis acquired in the pre-antibiotic era. They attended once a month, and were 'used' for teaching purposes, mainly to medical students and sometimes in postgraduate examinations. One such patient, anonymous to we junior doctors, had skull involvement. The bony part of his skull had been eaten away by a gumma, which is an inflammation caused by end stage syphilis. The medical students were asked to examine him. They often found nothing of note but had failed to ask him to take off his rather dapper trilby hat, which covered a 3 cm hole in his skull beneath which the outer brain covering, the dura, was visible.

Treatment with penicillin of course did not heal the breach. He had declined a metal prosthesis.

Bony remains from bygone times may show evidence of past syphilis. Finding such bones may help to resolve the origins of venereal syphilis, which has been passionately debated for centuries. There are three basic theories; the pre-Columbian, where syphilis existed in the Old World prior to Columbus discovering the New World, the Columbian theory which states that syphilis did not exist in the Old World until it was brought back by Columbus and his sailors and a combination of the two theories.

In 1994 a large collection of human remains was found when a site was excavated for a new Magistrates' Court in Kingston upon Hull. An archaeological dig ensued with subsequent discovery of a medieval Augustinian friary. Most of the hundreds of remains there were likely from the monks. However some were found to be wearing their 'Sunday Best'. These were likely to have been rich lay donors to the friary. Charlotte Roberts, a professor of paleopathology – the science of ascertaining what illnesses people were suffering from by examining their remains – has suggested that four of the skeletons in particular support the pre-Columbian theory. The usual method of dating skeletal remains is to use the radiocarbon (C^{14}) method. However, this could not be used in Hull because of the 'marine effect' caused by a high fish diet of the medieval subjects. The dating therefore relied on dendrochronology, that is dating of the wood that the coffins were made of, as well as stratigraphy, that is, estimating the date by looking at the layers wherein the remains were found. These methods dated the remains from between 1300 and 1450 – that is at least 42 years prior to the discovery of the New World. Examination of the skull and long bones showed a series of confluent pits and star-like radial scars as well as deposition of new bone. Microscopic examination also showed a combination of proliferative and destructive lesions. Roberts and her colleagues suggest the best explanation for these findings is therefore pre-Columbian syphilis. However they sounded a note of caution, saying that osteomyelitis, tuberculosis and leprosy might also have resulted in the findings. The Hull findings remain both exciting and controversial.

Goldmeier, D and Guallar, C. Syphilis: an update. Clinical Medicine (London) 2003 May–Jun;3(3):209–11.

von Hunnius, TE, Roberts, CA, Boylston, A and Saunders, SR. Histological identification of syphilis in pre-Columbian England. American Journal of Physical Anthropology 2006 Apr;129(4):559–66.

And so to the anus and rectum. Until the first occasion I was required to undertake a proctoscopy i.e. looking into the rectum, my thoughts on that part of the human anatomy were contained in a cerebral backwater. It was functional but non erotic. True, I had met that region as a medical student when vaginal tears at delivery extended into the anus and had to be carefully repaired by an experienced operator. And while doing my surgical house job the anus could be sealed off in severe cases of cancer and inflammation of the colon. And by this time I had two small sons, and was quite adept at changing their dirty nappies. Nevertheless most societies had relegated synonyms for the anorectal region to slang wording denoting something or someone contemptible. But I had no personal quarrel with the anorectum. I would give it the benefit of the doubt.

When I asked more experienced colleagues at the clinic as to how to do a proctoscopy, their answer was 'just put the instrument in'. I had followed that 'advice' and thought little about it, until considering how to be as accurate as I could while undertaking the research into the cause of proctitis (inflammation of the rectum) leading to an MD degree. I decided to go back to relearn the anatomy of the area. The word rectum means 'straight', and describes the three-dimensional route of the lower large bowel in the pelvis of mammals other than man. However in the human the rectum is both curved from front to back as well as from side to side. It straightens when the thighs are fully flexed, that is drawn up to the chest. That manoeuvre, and knowing the local topology made life kinder and easier for both the patient and myself. I had previously been taught that rectal infections in women could arise as secondary contamination from the vagina. But in homosexual men it was directly inserted at sex. It is nowadays well accepted that the anal area is an erogenous zone. Recent data shows that an increasing percent of heterosexual men and women have experienced anal sex over their sexual lifetimes, currently about 40 per cent for women and 50 per cent for men. And 90 per cent of gay men nowadays report having anal sex over the past year. But even in the mid 1970s, I learned from texts, anal sex could be pleasurable.

Tricia by this time was taking a five-year course in training to be a Kleinian child psychoanalyst. So, did our vocations have anything in common? Of course! The anus. Freud's second stage of psychosexual development from 18 months to 3 years detailed the anus as the primary erogenous zone, with the child deriving pleasure from controlling the bladder and bowels. Melanie Klein had later proposed that 'anal sadistic fantasies in which the object (and the self) is attacked by explosive faecal matter'. The knowledge that we might be both involved in anal pathology was contortedly and strangely comforting, even though in my heart of hearts I could only accept detailed psychoanalytical pronouncements as so much gibberish. But I did not divulge these thoughts to Tricia. She would learn in her own time how destructive Kleinian analysis could be. While looking at psychoanalytical literature I came across some pronouncements much closer to myself and the anus. You guessed it! Stammering. Freud had pontificated that stammerers 'project shitting onto speaking'. A later analyst Fenichel pronounced that a stammering child 'desires to withhold shit and thereby gain mastery of the external world, and at the same time wishes to spew the shit everywhere giving it pleasure'. After rejecting these fanciful ideas, I busied myself with the very here and now appearances and infection status of gay men's rectums.

My MD research work was supported by the Medical Research Council. An MD in the UK is the equivalent of a PhD but generally done over two years rather than three. It was an investigation into the cause of inflammation of the rectum of men having sex with men, in particular looking for CT. It took me 21 months to do the clinical work and almost three months to write up. There were no computers when I submitted it in 1976, so all 200 pages had to be typed using an electronic typewriter. I paid one of the medical secretaries called Sue to do some of the typing. This had to be done in the evenings. Tricia understood, but I feel was rather unhappy about me spending what often became long evenings with a young lady! Typing errors might mean either retyping the whole page or using correction fluid called Tippex to cover up the errors and type over them.

When completed, all 200 pages had to be photocopied three times. Photographs and data tables were reproduced in triplicate and pasted into the text. There were ultimately four bound copies. One hundred and thirty seven men having sex with men were entered into the study. As discussed above, there were no ethics committees to screen research work at that time. However, and in retrospect thankfully, I obtained verbal consent from each patient. The men were asked about rectal symptoms. I viewed the area through an instrument called a proctoscope. The image was magnified by use of an operating microscope (see picture).

Prah, P, Hickson, F and Bonell C, et al. Men who have sex with men in Great Britain: comparing methods and estimates from probability and convenience sample surveys. Sexually Transmitted Infections 2016 Sep;92(6):455–63.

Martin, D (2016) Stuttering from the Anus. Canadian Journal of Disability Studies, 5(3), 114–34 McBride, KR and Fortenberry, JD. Heterosexual anal sexuality and anal sex behaviors: a review. The Journal of Sex Research 2010 Mar;47(2):123–36.

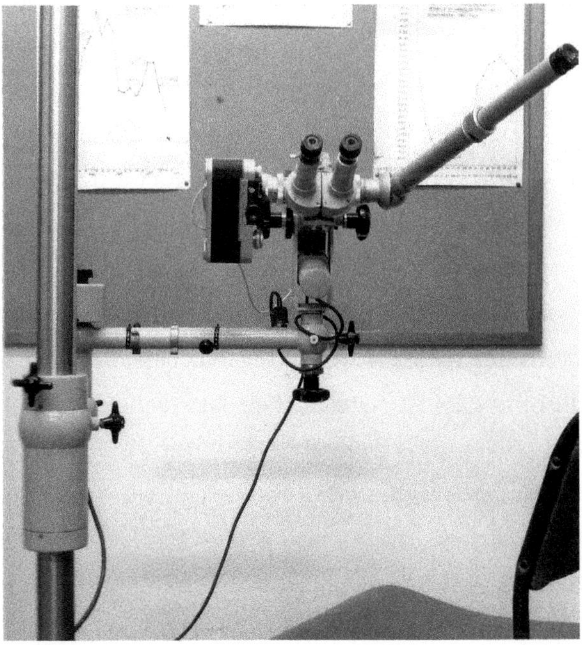

The operating microscope I used to view the anus and rectum

The testing for CT in the mid 1970s was done by infecting live cell cultures and looking for characteristic changes in the appearance of the cells. This work was undertaken by Dr S Darougar at the Institute of Ophthalmology in Judd Street London. The specimens were stored in a large container of liquid nitrogen that I transferred in my Austin Mini from Whitechapel to Judd street near St Pancras Station where the cells were infected. My research discovered two men who had CT in the rectum. I also found one man had the organism in his throat. Both of these were new discoveries. They were published in scientific journals as well as in the MD thesis. These short papers are never quoted nowadays. In order to find them in the literature one has to trace references from current papers on chlamydial proctitis back in time. In medical science, that which is novel and newsworthy for a few months is soon relegated to medical history. I retain two bound copies of my thesis. The third is now in the crypt of what was St Augustine and St Philip Church in Newark Street, which has become the Whitechapel site of the library of Queen Mary University.

In 1973 Tricia and I, plus our two small boys, moved house again, this time to a four-bedroomed detached house in Mill Hill, which we bought for £29,000 (today's value £1,300,000). We also were lucky to secure a wonderful Japanese au pair called Mizuko, who lovingly looked after our children, as well as making wonderful noodle soup and treating us, the parents, like demi-gods! When I arrived home from work she would gently bow forward with her hand palms together and with a sweet smile say, 'Good evening Dr Goldmeier. May I take your shoes?' Japanese noodle soup was always on offer from her. We were very fond of her and were sad to learn some 20 years later that she had died of ovarian cancer. Over the ensuing ten years we housed a succession of au pairs, some more memorable than others. Most were in their late teens or early twenties, some from the UK and others came from abroad to learn English. Edda Einesdottir was a tall thin Icelander who became a model and who taught me how to spin my children in a somersault. Wanita was a Seventh Day Adventist who became a mental health nurse. A young woman au pair from the UK seemed to eat a lot of bread, until we found out loaves were part of her bulimia regime. A rather overweight Italian au pair jammed my Mini car door against the pavement when she boarded the car. Another UK teenage au pair, in all innocence, found her way to the clinic I worked in, as she had an STD. We looked at each other in disbelief, as she sat in the waiting room and I called in the next patient (not her).

In 1974 I self-prescribed some codeine tablets for an episode of diarrhoea. It promptly curtailed the bowel problem but surprisingly also made me feel very happy and content. Alcohol, which I had tried in moderation over many years prior to this, served only to cause mild nausea and somnolence – so I mostly abstained. Coleridge initially used opium to manage toothache. His subsequent laudanum-produced euphoria was channelled into composing wonderful poetry, in particular 'Kubla Khan', also known as 'A vision in a Dream'. Addiction, whatever the substance used, finds its addict. So began my opiate problem which lasted 24 years. At first I took a small amount at weekends only. It seemed to expand my consciousness, perhaps by calming anxiety and allowing creativity to surface. Reading poetry, listening to music and appreciating art, took on new and expanded dimensions and vistas. I did not take it while at work, at least not in those early years. The dreamy state it induced would be supressed by logical thought processes, which themselves would be impaired by the opiate. I watched popular television programmes of the times while under the influence of what was initially a small dose of codeine. These included music from the *Kojak* series with Telly Savalas, and Rod Stewart singing 'I am sailing' in a series about the 'Ark Royal' aircraft carrier. If I hear even a few bars of their music nowadays, they conjure up a shadow, a silhouette, a reflection, if you like, of opiate euphoria and how I felt over 50 years previously. This is a great example of Pavlovian conditioning, where the music is the conditioned stimulus and the codeine the unconditional stimulus that I last took almost 27 years before. Such is the strong brain bond between opiates and the brain. It never goes away! In those early drug using days I rationalised weekend use as no different from drinking small amounts of alcohol. Whatever the drug of choice, the early stages of addiction are a love affair. The codeine pleased me whenever we met. As time progressed I required higher doses to produce a euphoric feeling. Later still it had to be taken so as not to go into opiate withdrawal. But by then it had become a despoiling ogre.

By 1975 we had a family of three sons aged from four months to four years old. We decided to go for two weeks that summer to Lelant in Cornwall, where we stayed in a small vegan hotel. We used the then still popular Motorail to transport the car from London to St Austell, from where it was a relatively short drive to Carbis Bay and Lelant. A French au pair who was part of our family at the time expressed daily disappointment when presented with the hotel's bland culinary fare – nut loaf at every meal with a tasteless brown sauce, boiled potatoes, vegetables steamed or boiled with no added seasoning. It was a monotonous diet fit for a puritan. As if to compensate, the north Cornish beaches were pleasing to the eye, with their large expanses of flat sand leading to the Celtic Sea. While my older two boys played with bucket and spade, I gazed at the very blue sky, produced by the reflection of the light off granite sand. It was an intoxicating mix of excitement and tranquillity. High flying civil aircraft appeared as silver darts in the sky. I had space and time to compose a poem and later to paint a fantasised picture of what I had seen.

I loved you first at golden Cornwall beach
When I was young and plain with wife and children
Celestial dagger ripping white the Columbia blue
You disappeared, whispering with a roar
Into the cobalt of the Atlantic
But most of all you are my bosom friend
Just behind my very own suburban house
Where the olive tree, rose, gorse heathers and old garden shed
Condense to finderscope the city's Celeste blue
Behind and above and to your sides

I am you
In this moment
No time
No space
Bubbled in hope
Intense and vast
Jubilant and joyous

Volte face – I'm up there high
'Mid the fliers in the sky
Is my seat 10A or C?
Is the forward toilet free?
Why the cabin crew so old?
Why the filtered air so cold?
Why the child in seat 9C?
Is screaming loud to do a wee?
Can the pilot steer the plane?
Are the engines under strain?
Will the plastic food arrive?
Is the touch screen video live?
Why do the wings rock to and fro?

Why are the landing lights aglow?
Do the LACC at Swanwick know?
Where we are – and does it show?

Serenity regained
I lie on my beach blanket
And watch your white contrail edge
South to North
Up to bonny Scotland
Where you will chase the setting sun
And I am content in this one cerulean moment.

Chapter 8

PSYCHIATRY, GENERAL PRACTICE AND GU MEDICINE

Soon after I had gained my MD qualification I made a decision to obtain some experience in psychiatry. This in part was to dip my toe into the field of psychological health to see if that speciality suited me, and partly to consider the possibility of working as a more psychologically minded venereologist. Tricia by this time was well into her child psychoanalysis training at the Tavistock clinic, so both of us were so to speak becoming more psychologically minded.

The best place in the UK to train in psychiatry at that time was the Maudsley Hospital in Denmark Hill and its ancillary units. I was accepted first time round so I guess the interview went well. I started my first rotation at the Unit for Psychiatry of Older Age at the Bethlem Royal Hospital in Beckenham Kent, working for Professor Raymond Levy. The Bethlem was originally founded in 1274 and formed into England's first hospital for 'the insane' in 1377. It became infamous in later centuries as a venue for folk to visit the hospital to be 'entertained' by viewing the mentally sick inmates. I travelled to The Bethlem in Kent in my Mini – something that would take twice as long now because of 21st-century traffic, to say nothing about passing through the congestion and emission zones. We assessed each patient in great depth, and so had only a handful under our care. The ward rounds were very academic, and could go on for hours discussing many different features of just one patient, their possible diagnoses and related relevant research papers. Professor Levy had an encyclopaedic knowledge of psychiatry in general and was a pleasant, quiet spoken man who occasionally took snuff while expounding. He went on to write many papers and books, some in French as his formative years were spent in Cairo, including a book *Mamnou'*, which apparently means flexible and positive about life, in which he describes his Cairo upbringing, where he acted in a school play alongside Omar Sharif. Again, I found myself becoming very fearful presenting cases because of my stammer, my speech almost grinding to a halt. That said, all the members of the team listening to me were very patient. Psychiatry involves a great amount of verbal communication in order to diagnose and manage patients. A patient with acute appendicitis will have maybe a five-minute history taken by a surgeon, who will then examine her to make the diagnosis. Acute crushing chest pain will lead to a very brief history and a diagnostic ECG to confirm a heart attack. On the other hand a psychiatrist will need to ask questions relating to personality, appearance, speech, thought, thought content, abnormal beliefs, insight into illness and in older people cognitive testing in order to assess the risk of dementia. Once the whole picture is obtained it has to be retold at a ward round. A lot of speaking required! Compare that to showing a colleague a diagnostic ECG. I decided after about six months that I could not endure the high level of anxiety. Finally I capitulated and told Prof Levy that the journey was too far and that my oldest son Oliver was having bad asthma that needed hospital admission often – that bit was true. During my time at Bethlem I took codeine on many occasions during the day.

I had to think fast. What job to do next?

I looked at the job advertisements in the *British Medical Journal* and found one as a trainee GP in Elstree. I was interviewed by a senior partner Henry Collins, and got the job. I purchased a large black GP case to carry medications that I would need treating patients seen at home. One of the other partners was Dr John Marks, who went on to be chairman of the British Medical Association, and was an inspiring teacher. The practice saw mainly NHS patients, although there was some private that the partners practised which included the Elstree film studios. I was tasked to assess none other than Rudolf Nureyev, who was starring in Ken Russell's film *Valentino*. The studio had contacted the practice informing the partners that their leading man was indisposed on set because of a fever and malaise. I am still unclear why I was sent. Nevertheless as a dutiful trainee GP I was keen to help the patient. As I approached the film star, someone called out, 'the doctor has arrived'. Nureyev glanced at me momentarily, and with a flick of his head and a wave of his hand dismissed me, saying he wanted to see his own doctor. So he never became my patient. My memory of him is of a petulant, smelly, sweaty prima donna or rather prima uomo, to use the correct term for a man.

Home visits were not uncommonly undertaken in the mid 1970s. I diagnosed a heart attack in a 60-year-old man, based on his ECG. Dr Marks told me as the patient was otherwise well that he could be treated at home. Of course nowadays he would be given an emergency angiogram, and have a coronary stent inserted. I attended him daily, but was told one morning that he had died the night before. I issued the family with a death certificate after seeing the body in pyjamas. Managing terminally ill patients was routine. I attended a dying man who was surrounded by his family, wife and girlfriend who all were there for him in unison. I prescribed Brompton's cocktail to ease his suffering. It contained morphine plus cocaine and chlorpromazine. The local chemist was expert at compounding this mixture into syrup form. Treating sick children at evenings or weekends (the practice shared on calls with other local practices, so that any one practice was on call one in three) could be worrying. On one such home visit, I initiated treatment for a three-year-old boy who had a high fever, pus covered tonsils and who really looked unwell. I took out a bottle of antibiotic powder from my case and added water to produce the syrup he needed. I was worried about the child, so I sat by him after he took the medication. Much to my and the parents' distress and dismay, he promptly vomited up the lot! To my hospital doctor mind, this event could only be followed by hospital admission and intravenous treatment. I conveyed this sorry sequence of events to Dr Marks on the phone, merely to set the scene for his hospital admission. Instead of hearing something like 'it does sound like admission is what he needs' he bellowed, 'Not bloody likely'. He instructed me to sit with the child and spoon feed him very small volumes every 10 minutes until 10 ml had been ingested. And it worked – he kept it down! I left the parents an hour later but returned the next day to find a happy three year old and even happier parents. Such was the acme of general practice that I was taught by John Marks.

A final case worth mentioning was a woman who had called the local council hygiene department a number of times to have insects in her house eradicated. The pest controllers could find nothing time after time. I was asked to assess her. She pencil drew large ladybird like creatures with a decorative noughts and crosses pattern on their on their backs. Her visual hallucinations turned out to be the presenting feature of an early dementing illness.

Talking on the phone (of course they were all the patients' landlines in those days) when referring patients to hospital was again anxiety provoking for me, particularly when describing clinical features that were being questioned by the hospital doctors. In the end I stayed in Elstree for eight months. I might have made a career of general practice and Dr Marks encouraged me to do so. But I felt I ought to have another try at psychiatry.

I took my little family to Weymouth in summer 1976. I interrupted the vacation to go for interview at the Royal Free Hospital Rotation in psychiatry which I started at the Royal Free Hospital itself, working for Drs Don Hailstone and Nancy Swift. Dr Hailstone was obsessed with sailing, and spent most free weekends at Portsmouth where his ocean going yacht was moored. Dr Swift combined talking psychotherapy with psychiatry and its medication treatments, and was held in high regard by her colleagues. Her husband had died in action in 1943. She never remarried. She supervised me while I regularly saw a novice nun who had lost her voice, likely not due to physical causes. Dr Swift, smoking her cigarette in a long black holder, told me in her calm, rather croaky voice, 'Probably something sexual David that you need to find out'. I never did find out or indeed help the nun. There was a regional anorexia unit in the department headed by Professor Gerald Russell. I presented a case at an internal conference he was chairing after which he told me I had done well. He also contacted me one weekend when I was on call to ask me apologetically to prescribe pethidine for his wife's back pain, which I duly did – but did find it rather odd that he could not go via his GP! But by this time my fear of speaking and generalised anxiety was getting the better of me. I felt I could not think or remember patient's details. I was slipping into depression. My mother found my mood hard to cope with and became very distressed. Somehow I was able to discuss my situation with another psychiatrist at the Royal Free, Dr Silvio Benaim who was well known in the Royal Free having diagnosed an outbreak of mass hysteria among the nursing staff some years earlier. He arranged for me to be seen and admitted to Guy's Hospital under care of Dr Jack Fleminger. I was diagnosed as having depression and anxiety and was given a single room on the psychiatric wards. I was isolated for the length of my stay. No psychological intervention. No occupational therapy. No interaction with other patients (maybe that was a plus point?), just waiting for the medication to work. However I ran once or twice a day over London Bridge and into the City and have continued almost daily cardio exercise ever since. For me, cardio exercise most definitely improves my mood, lowers anxiety and facilitates thinking and creativity.

Regular cardio exercise has both short and long term benefits on the central nervous system and mood. Mild to moderate depression is improved in patients who undertake regular exercise. Chronic exercise also increases blood flow to the brain and encourages the formation of new nerve cells and their connections. The 'high' or euphoric feeling in the hour or two after cardio exercise is likely underpinned by an increase in neurotransmitters, including opiates (endorphins), adrenaline, serotonin and dopamine.

Martinsen, EW, Medhus, A and Sandvik, L. Effects of aerobic exercise on depression: a controlled study. British Medical Journal (Clinical Research Edition). 1985 Jul 13;291(6488):109.

Gligoroska, JP and Manchevska, S. The effect of physical activity on cognition – physiological mechanisms. Materia Socio Medica 2012;24(3):198–202.

After discharge from the hospital my daily running route was from my house in Mill Hill to the Graham White RAF museum in Hendon.

Turn the warm and cosy covers
For the shaky early start
Search the room for ever hover
Where the shoes to race my heart?

Green the shirt and blue the tracksuit
Cover over spindly legs
'Eat your meat and leave no veg'
Mother used to chide and beg

Leave behind those many mem'ries
Sticky, strong they tug away
Brave the Sunday frost unfriendly
Gosh it's cold at break of day

Softly now I shut the portal
Slumber little family
Solitary, shivering mortal
Jogging into harmony

To the RAF at Hendon
One point five the miles each way
Oh my hopes and oh my tendons
Rush towards the light of day

Turbo props and 747s
Seem to gather overhead
Landing lights adorn the heavens
Wish I too some wings could spread

Workhorse sparkling in the sunshine
Perfect work is nearly done
Now descending exhausted shrill cries
Farewell stars and moon and sun

Jog along the tree-lined avenue
Branches arching overhead
Graham White the halfway venue
Beverley transporter wings outspread

Turn about sunrise behind me
Tic tac tic tac the train transcends
Hissing steam and smoke through chimney
Or electric, cabs each end

Sweating skin and sinews straining
Homewards now I bound with glee
Not long now the puffing panting
Soon relax with mug of tea

Hurry home oh how I hunger
Take a deep breath comfortably
But now no thoughts nor mem'ries, anger
Just my body paining me

Running running ever running
Will this jogging never cease
Till the last day ever coming
Buried snug my mind at peace.

I never did see Dr Fleminger. Rather his pleasant senior registrar Dr Ros Furlong came to discuss my progress on a regular basis. She became a consultant at Friern Hospital, where I was about to work. I was put onto antidepressants and lorazepam. Lorazepam is a benzodiazepine type medication, which was very popular around 1977, as it and other drugs in its class were not lethal in overdose as were the barbiturates, which had been previously used to treat anxiety. Some ten years later new data was published showing that they induced dependency when taken regularly for more than four weeks, and that coming off them produced intense and debilitating side effects that could linger for up to a year. I came off the lorazepam over a period of a year about six years later. My blood pressure again was found to be very high in Guy's but came down with my psychotropic medication, so that no specific treatment for that was given. Tricia came to visit me a few times, as well as my parents and my brother (who always found talking about feelings immensely trying). I was discharged after a few weeks. I am not sure my depression had improved but the anxiety was certainly better. I had not told any of my psychiatric carers about taking codeine, and continued to take it at weekends. I was both ashamed of myself for not reporting this, and at the same time relieved that I could continue to enjoy what I perceived to be precious moments of relaxation. Such is the insanity of addiction, but more of that to follow. My resumption of work was at Friern Hospital, which was part of the Royal Free Rotation. Friern Hospital was built near where Southgate and Whetstone are situated nowadays, close by a hamlet named Colney Hatch. For much of the 20th century its former name, Colney Hatch Lunatic Asylum, was synonymous among Londoners as the archetypal mental institution.

The name Colney Hatch appears in CS Lewis's Narnia chronicles to describe the empress Jadis's apparent disordered mental state as she cavorts through London, attacking people, robbing a jewellery shop and riding on top of a horse drawn cab as if it were a chariot. She then states, 'We are the Empress Jadis.' 'Ho! Hempress, are you? We'll see about that,' said a voice. Then another voice said, 'Three cheers for the Hempress of Colney Hatch'.

It also appears in a Jeeves and Wooster book. 'I nestled against the pillows, the brow a bit drawn. It is all very well to try to do old school friends a spot of good, but I could not but feel that in espousing the cause of a lunkhead capable of mucking things up as Gussie had done, I had taken on a contract almost too big for human consumption. It seemed to me that what Gussie needed was not so much the advice of a seasoned man of the world as a padded cell in Colney Hatch and a couple of good keepers to see that he did not set the place on fire.'

It was opened in 1849 by Prince Albert and at its height housed 2,500 mentally sick patients. The estate had its own water supply, chapel, cemetery and 75 acre farm. Fire destroyed some of the building in 1903. By 1927 there were nearly 600 staff employed on the site. It was renamed Friern Hospital in 1958, having been incorporated into the NHS in 1948 from its former controller the London County Council. In spite of adding Halliwick House as an acute ward in 1958, Friern's days were numbered and it closed in 1993, when many of the patients were 'integrated' into the community. None other than Jeremy Corbyn, as a local MP, raised the issue of the impending closure of Friern in Parliament in 1990, adding that not enough money or service provision had been earmarked for the patients who were to be decanted into the district from the acute wards. I thought the Italianate style of the Friern building ugly. The back wards housed many hundreds of patients who had been there for decades and had become institutionalised. It was not uncommon to find single women in their 70s or 80s who had developed puerperal depression in their youth and had lingered on as lost, forgotten souls into old age. Their babies had been taken away from them immediately after birth. Mostly they never met their children again. Being on call to visit sick long-term inmates meant negotiating long serpentine, smelly, dark gloomy corridors and opening the ward entrance with a large Victorian key. The building has since been converted into luxury flats.

At Friern I worked for Dr Doris Hollander, a consultant psychiatrist. Her halo of dark brown frizzy hair belied her tough yet compassionate approach to patient care. She always addressed me as 'Dr Goldmeier' in her light South Africa accent, rather than my first name. I found out later that she was born in 1936 in Germany and had escaped the Nazis with her family, via Holland, to southern Africa. Her career over the years was certainly not conventional. She fought many battles against apartheid in South Africa and temporarily left her training post at the Maudsley to help in Israel during the Six-Day War. Ward rounds at Friern were still troublesome for me in terms of me speaking but I managed. The team included a clinical psychologist Dr Maureen Sheehan, and a male psychiatric social worker whose name I cannot recall. He had been a ballet dancer at some earlier point in his life and when the he felt even a touch of tedium (often) his thoughts wandered from the proceedings of the ward round meeting and entered a ballet reverie, where he would recount the exact steps of ballets and movements in sequence. His favourite was the Pas de Deux from *Swan Lake*. An irate Doris Hollander would angrily interrupt these reveries.

RELATIVES AND FRIENDS GROUP IN A PSYCHIATRIC HOSPITAL

The ward I worked on in Friern was, like many in that establishment, high ceilinged, dark, dreary and sinister. An aside here. Sinister derives from the Latin meaning left sided, implying danger or harmful. Dexterous derives from the Latin referring to the right side but also implying being quick witted or skilled. Society in the past was highly prejudiced against we, the left handed people. But let us return to Friern. The acute ward was filled with the depressed, manic and schizophrenic. Many lived out their youth, adulthood and middle age years in the dullness of Friern, their time peppered with attempts, often short lived and unsuccessful, at discharge into the community. I only remember the name of one patient under my care there from that time – Bellerose. She was in her late thirties and wore clothes of various shades of brown day in and day out, which matched her olive skin and black frizzy hair. She never seemed to be in the queue at medication rounds. That being so, the high ceilinged ward often echoed to her name, now quieter, now louder, as the charge nurse roamed the multitudinous nooks and crannies of the maze of rooms in search of their patient. She was a chronic depressive, seemingly unresponsive to treatment. One day, when I guess the wanderlust had failed her, I found her sitting in the capacious dayroom. She was staring at the windows, where a grey squirrel had emerged into the pallid winter sunshine and sat motionless looking at both of us as if it required our attention. As I walked on I heard a whisper – 'My father wants to know why I can't get better'. I looked round the room but there were only the two of us – so it must have been Bellerose addressing me. I asked if she had spoken to the charge nurse about her request. She had not. She went on to tell me that there were other medical and social issues that her father wanted to discuss. 'Leave it with me,' I rather rashly proclaimed. I asked a few other patients if their friends and family had been given the opportunity to discuss their progress with the staff. In the main the answer was 'No'.

At the next ward round I put the forward the idea of a group for the friends and family of our patients to Dr Hollander and Maureen Sheehan. They were somewhat taken aback by the idea, but after brief discussion thought it an interesting and worthwhile project. The nursing staff and psychiatric social worker were also in agreement. We then took the idea to the 30 patients in the unit and asked how they felt about their family and friends meeting to talk about and discuss them. They were very positive and even excited about our proposition. We therefore set up the meeting in the day room from 7 to 8 pm each Wednesday evening. The patients knew the meeting was ongoing but did not take part. In all, 50 relatives and friends attended over a 32 week period. They initiated the discussion rather than any of the staff. Positive outcomes of the meeting were that the family and friends could meet the staff and that it was a forum for questions about the patients' illnesses and treatments. The relatives and friends further reported that the meetings made them feel more supported, often gaining this from each other rather than from the staff who tended to input more about practical issues. The staff felt that they gained more insight into the patients' background via the group. I left the unit before the group ended and am not sure if it continued after the 32 weeks. We wrote up our findings in the British Medical Journal.

Goldmeier, D, Hollander, D and Sheehan MJ. Relatives and friends group in a psychiatric ward. British Medical Journal 1979 Apr 7;1(6168):932–4.

Later on in the job, Maureen and I teamed up to start a once a week conjoint sexual dysfunction clinic at Halliwick. At that time sexual dysfunction was deemed to be largely psychological in aetiology with publication of the seminal works of Masters and Johnson leading the way.

William Masters, who trained as a gynaecologist, met Virginia Johnson, a former singer and sociologist in the 1950s USA. From then onwards they did research on '10,000 sexual cycles' and produced a book Human Sexual Response, *which debunked many myths about sexual responses e.g. that vaginal lubrication came from the cervix (they saw it appear in droplets on the vaginal wall of women in states of arousal). They went on to author a book* Human Sexual Inadequacy *which described what they considered to be the causes of sexual problems, as well as explanations as to how to manage these problems, which they felt in the main were psychological in aetiology. Masters left his first wife to marry Johnson, but later divorced her too. Masters and Johnson failed to describe a desire phase in their human sexual response. That was left to Dr Helen Singer Kaplan who added it in 1979. Masters and Johnson also undertook treatment to 'convert' homosexual men's sexuality.*

I met William Masters briefly at a conference in the USA, where he and I shared an elevator. He had severe Parkinson's disease by this time and was wheelchair bound. His third wife, resplendent with purple hair, pushed him around the conference halls, where he was universally lauded and feted.

Pines M. 'Human Sexual Response' – A discussion of the work of Masters and Johnson. Journal of Psychosomatic Research 1968 Jun;12(1):39–49.

The backbone of our treatment was sensate focus, as had initially been described by Masters and Johnson. That involved putting a ban on sex or even attempting it, and instructing the patients to take it in turns to gently massage each other in non-judgemental fashion, at first choosing non genital areas and later on focusing on the genitals, but with instructions to notice the sensations that are present and to try not to get involved as to whether the sexual organs are functioning or not. Although we did not appreciate it at the time we were in effect using mindfulness techniques. Another useful ploy was to give patients paradoxical instructions e.g. men with psychogenic erectile dysfunction (ED) (we wrongly presumed at that time that all ED was psychogenic!) were instructed to try as hard as they could NOT to get erect. In some cases of psychogenic ED it worked very well. It was only in 1983 that Giles Brindley, the eccentric but brilliant UK physiologist, gave his evening lecture to a urological conference in Las Vegas where he demonstrated the erection he had produced by prior self injection with phenoxybenzamine (an alpha adrenergic blocker which dilates blood vessels). Having pulled down his tracksuit trousers and underpants, he felt the audience did not get a good enough view. He therefore waddled with trousers and pants at his ankles down from the podium to give the audience a better view, in so doing shocking some of the male urologists' wives. Brindley was an eccentric polymath physician physiologist, who seemed to revel in demonstrating the workings of the body on himself. At one undergraduate teaching session he entered, he put a padded cushion on the table, proceeded to stand on his head and drink a glass of water, so demonstrating that food and drink pass down the oesophagus largely because of contraction of that organ (peristalsis), not by gravity.

I continued to run the clinic with Maureen after I left the Royal Free rotation as I was given an honorary contract by the hospital.

While still at Friern I became marginally involved with a patient under care of Dr D'Orban a forensic psychiatrist. His patient had been a journalist who became very depressed on his way by train from London to a conference in Blackpool. He turned back convinced that the world for him was a black and depressing place. He told the staff he had wanted to end his life and take that of his Japanese wife with him. He drowned her in their bath and then took an overdose of tablets. His life was saved by a neighbour. At his trial he was convicted of murdering his wife, but sentenced to treatment for depression rather than serving a jail sentence. Many of my colleagues at the time felt he had literally got away with murder! He was discharged some months later on, no longer being depressed and a free man.

In 1978 I made a decision to move back to Venereology and so applied to work at University College Hospital London (UCLH). I guess I was pretty well qualified for the senior registrar job there. I interviewed reasonably well and so worked for Dr David Oriel, who had been a consultant there for a number of years, having been in the RAF and then doing some work in general practice prior to Venereology. He was a tall, quietly spoken, somewhat hard of hearing, bright and very organised man with a generous smile, who knew his subject and had written many papers on genital warts and chlamydia. His chlamydial research co-worker was Dr Geoff Ridgway who was the senior medical microbiologist at UCLH. They jointly authored many papers. Dr Oriel was in great demand as a speaker and was well known and respected in international circles, particularly the USA. He could lecture in French and in those days of 35 mm slides as aids to talks, was able to show two sets at once to better illustrate any point he was making. In terms of treatment he used many more sensible USA regimes such as one injection of benzathine penicillin for early syphilis rather than daily procaine penicillin as was generally used in the UK at that time.

PROCAINE PENICILLIN

In 1980, when working at UCLH, I undertook a clinical trial to ascertain the level of penicillin that could be found in blood and cerebrospinal fluid (CSF) – that is the fluid around the brain and spinal cord – in patients treated for syphilis. The tenth patient in the study was Jacob, a burly 27-year-old carpenter who was found to have positive syphilis blood tests. He had come to the clinic because of a urethral discharge which was found to be caused by Chlamydia trachomatis. As per routine protocol, we repeated the blood tests for syphilis to confirm the initial result. With both tests positive we were able to stage his syphilis as late latent, which meant he was no longer infectious and had passed the primary stage (usually genital ulcer) and secondary stage (usually a rash, generalised lymph node enlargement plus other features) but had not yet entered the third or tertiary stage where bones, the heart or nervous system are involved. Latent syphilis needs to be treated to prevent progression to the tertiary stage. Pure penicillin is rapidly absorbed and excreted from the body, meaning that if blood levels are to be active throughout the day it must be injected four times a day. What is needed is a once a day or better still once a week injection where the penicillin is stored in the body (repository) and released over time but still giving effective penicillin levels all day long. Nowadays this is with an intramuscular injection of benzathine penicillin. Once injected into the buttocks, this repository antibiotic releases penicillin gradually over a week. We were unable to find this product in 1980, so we used a shorter acting repository penicillin called procaine penicillin that was thought to give good levels in the blood for just over a day. Procaine is a local anaesthetic. It combines with penicillin by exchange of ions to produce crystals which dissolve well in peanut oil. I was present when the treatments were given to all the patients in the study. The daily injections were given by nurses who were much more adept than I at administering the intramuscular injection into the upper outer quadrant of the buttock, so as to not harm the sciatic nerve. In Jacob's case, half a minute after the injection had been given, he became visibly shaky, turned to me with glaring eyes and said, 'you shit, you just killed me'. He lunged towards me. However he was intercepted by a powerfully built male charge nurse who restrained him gently but firmly. Jacob calmed a minute or two later. His response to the injection is termed the procaine reaction, whereby the end of the needle is inadvertently inside a vein, so that the drug is effectively given intravenously. It is the procaine part of the drug that causes a temporary psychotic reaction, when a large volume of it reaches the brain. Sucking back on the syringe before injecting can indicate that one is in a vein, when blood appears – however sometimes when sucking back the vein just collapses and no blood is seen.

All his subsequent injections were uneventful – he had 14 in all. Twelve hours after the second injection I performed a lumbar puncture, that is putting a needle into the area around his spinal cord to obtain CSF and at the same time took a specimen of blood. We had no facility for doing this at UCLH, so all patients went to the National Temperance Hospital in Camden. As the name suggests, the hospital strongly discouraged the use of alcohol in the treatment of patients. It was opened in 1873 under the initiative of the National Temperance League but was transferred to the NHS for general use in 1948. It closed in 1990 and was later demolished to make way for the High Speed rail line out of Euston station.

Penicillin G is the naturally occurring penicillin that is derived from the basic fermentation process of the penicillin mould. It was used to treat syphilis as early as the latter part of World War 2. In the very early days of penicillin, the dose required in order to result in a successful treatment outcome had yet to be determined. Rabbits were therefore inoculated with the causative agent of syphilis, Treponema pallidum, which then caused an ulcer. The testes were the usual inoculation site. The dose per unit weight of the rabbit and matching blood penicillin level that resulted in cure of the ulcer were determined. The World Health Organization determined that a level of at least 0.018 ug/ml of penicillin was required over at least a week to ensure a clinical cure. Higher levels are likely to work faster. In the 18 patients in my study, four different dosages of procaine penicillin were used. All gave a level greater than 0.018 ug/ml in blood but none reached this level in the CSF. Routine benzathine penicillin given nowadays gives similar results. In spite of this absence of antibiotic in the fluid encasing the brain, patients once treated with routine dosages of procaine or benzathine penicillin never go on to get syphilis of the nervous system or heart. This is somewhat of an enigma! One mechanism whereby these seemingly low CSF penicillin levels prevent progression to late disease may be because blood levels are more important than those in the CSF.

I undertook this research work with Pamela Waterworth, a microbiologist who measured the penicillin levels against bacteria of known sensitivity to the agent. She had worked as a microbiologist at St Bartholomew's Hospital from 1943, where she undertook research with the Penicillin Clinical Trials committee in the early days of antibiotic use in the UK. She moved to the Hammersmith Hospital in 1961 where she held the post of senior demonstrator in chemotherapy and finally University College Hospital where I worked with her.

The story of how penicillin was initially produced on an industrial scale during World War 2 is always worth repeating. The UK was unable to undertake this, so the baton was passed to the USA. Problems of a poor initial growth of the penicillin mould were overcome when a strong growth of it was found on a rotting cantaloupe melon in a street market in 1943. This was then transferred to large vats containing corn steep liquor (a sugary broth made from maize) in a commercial process in Preoria in Illinois.

Liggett, RW and Koffler, H. Corn Steep Liquor in Microbiology. Bacteriological Reviews 1948 Dec;12(4):297–311.

Arseculeratne, SN and Arseculeratne, G. A re-appraisal of the conventional history of antibiosis and Penicillin. Mycoses. 2017 May;60(5):343–347,

Goldmeier, D and Waterworth, PM. Penetration of penicillin into the cerebrospinal fluid of patients with latent syphilis. Pharmatherapeutica. 1981;3(1):14–7. PMID: 7323106.

Young, MY, Andrews, GW and Montgomery, DM. Procaine penicillin. The Lancet. 1949 May 21;1(6560):863–5.

David Oriel checked all the notes of patients seen the previous day prior to commencing work the next day, and so had his finger on the pulse of what was going on in his department. If he perceived I had made an error he told me so in constructive fashion. My own interest lay in the psychological aspect of patient care, and in 1980 I undertook the first of a number of studies on that aspect of genital herpes.

In 1982 I wrote an article with Anthony Johnson of the Medical Research Council biostatistics unit in Cambridge, on the possible effect of depression and anxiety on the recurrence rate of genital herpes. Patients who had their first outbreak of genital herpes were followed up until they either returned and could have their first recurrence confirmed by virus testing, or if they did not return, asking them by post when their first recurrence outbreak had occurred. All patients were given an inventory called the General Health Questionnaire (GHQ). Those who scored higher were much more likely to have significant anxiety and/or depression. Although there were a number of methodological problems with the study, it did seem to demonstrate that a higher GHQ score when first seen predicted a more rapid recurrence. It was, and still is unclear if such psychiatric illness pre-exists and results in more frequent herpes recurrences, or rather the psychiatric illness is the result of the stress of having the infection.

A further research paper I wrote in 1988 seemed to confirm the association between non psychotic psychiatric illness and recurrences of genital herpes.

I tried as hard as I could to improve the lives of patients who were distressed by their recurrent genital herpes patient. This included medical hypnosis, to suggest a more positive outlook on life. I never did document scientifically if this treatment approach helped. However after hypnotising a junior colleague who worked with me in the herpes clinic, Dr Simon Barton, suggesting that he would have a high batting cricket score in an oncoming match, he in fact scored 100 runs at the crease. Make of that what you will!!

Simon Barton in later years went on to hold a personal chair in what was to become Genito Urinary Medicine at Imperial College. He and his team undertook a more detailed study of the effect of medical hypnosis on patients suffering from frequently recurrent genital herpes. The patients were assessed six weeks before and again six weeks after the hypnosis. Overall there was a decrease in the number of patient outbreaks after the hypnosis, along with an increase in CD3 and CD8 lymphocytes, which generally aid in immune defence. Those patients who improved the most had decreased anxiety and increased natural killer cell count.

In later years I wrote two more relevant papers, in particular looking at how the immune system interacts with the virus and the patient's psychological status.

Most patients with recurrent genital herpes cope with it very well and are able to disclose their infection to prospective new partners with relative ease. A minority, even nowadays, feel tainted, mentally scarred and need a great deal of compassionate psychological care. Some find it impossible to disclose their infection to a new partner. In 2011 the Court of Appeal in England upheld the conviction of a jail sentence for David Golding, who the lower courts deemed had recklessly infected a sexual partner by not informing her of his genital herpes prior to sex, and not taking any precautions to prevent infecting her.

In the early 2000s Matthew Hamill and I undertook a survey of STI clinics in the UK to determine which units offered counselling for patients with genital herpes. The good news was that two thirds of those clinics responding provided such facilities. After leaving my department Matthew studied anthropology at Oxford and proceeded to do a PhD at Cambridge. He now works at Johns Hopkins in the USA.

Goldmeier, D. Herpetic proctitis and sacral radiculomyelopathy in homosexual men. British Medical Journal 1979 Sep 1;2(6189):549.

Goldmeier, D and Johnson, A. Does psychiatric illness affect the recurrence rate of genital herpes? British Journal of Venereal Diseases 1982 Feb;58(1):40–3.

Goldmeier, D, Johnson, A and Jeffries, D et al. Psychological aspects of recurrences of genital herpes. Journal of Psychosomatic Research. 1986;30(5):601–8.

Goldmeier, D, Johnson, A, Byrne, M and Barton, S. Psychosocial implications of recurrent genital herpes simplex virus infection. Genitourinary Medicine 1988 Oct;64(5):327–30.

Goldmeier, D. Genital herpes: Heisenberg revisited. Sexually Transmitted Infections 1998 Jun;74(3):219–20.

Hamill, M and Goldmeier, D. Management of recurrent genital herpes: a survey of UK genitourinary medicine clinics in 2003. International Journal of STD & AIDS. 2005 Nov;16(11):760–2.

Goldmeier, D, Garvey, L and Barton S. Does chronic stress lead to increased rates of recurrences of genital herpes – a review of the psychoneuroimmunological evidence? International Journal of STD & AIDS. 2008 Jun;19(6):359–62.

Fox, PA, Henderson, DC, Barton, SE, Champion, AJ, Rollin, MS, Catalan. J, McCormack, SM and Gruzelier, J. Immunological markers of frequently recurrent genital herpes simplex virus and their response to hypnotherapy: a pilot study. International Journal of STD & AIDS. 1999 Nov;10(11):730–4.

While at UCLH I also took on a patient for brief analytical psychotherapy, seeing her weekly, for an hour. This type of therapy is a form of psychoanalysis. It focuses on dialogue between the therapist and client, where unconscious elements relating to repressed fears and conflicts are brought into the conscious mind by techniques such as free association and dream analysis. A much held viewpoint is that such therapies are not effective, although they were highly popular in New York during the 20th century, but not nearly so favoured in the UK, which on the whole used cognitive behavioural therapy was used to manage psychological problems such as depression and anxiety. Woody Allen reported that 13 years of classical analysis had 'helped him to grow older' but maybe not much else! Some critics feel it is the warmth and ingenuity of the therapist rather than any technique he or she uses, that results in progress. Peter Fonagy, who fits the bill for these therapist's characteristics, presented objective data showing psychotherapy could indeed help some patients with anxiety and depression. There were reasons other than to gain the experience that pushed me to undertake this work. I believe I was better able to understand what Tricia was going through with her personal Kleinian analysis, five days a week, which was a requirement for her training as a child psychoanalyst. I and five UCLH medical students had monthly supervision for the therapy we were providing which was led by the psychotherapist Irene Bloomfield. Irene was a refugee from Nazi Germany. At the age of 14 she refused to give the Heil Hitler sign when asked to do so, and walked past a Sturmtruppen (Stormtrooper) who was preventing people going into Jewish shops in Berlin. She eventually came to the UK on a domestic's permit (as my mother did!) just prior to World War 2 and went on to develop the Raphael Centre for counselling for Jewish people.

I think Dr Oriel was sympathetic yet remained somewhat suspicious and uncomprehending of my psychological interests.

Part of my week was spent at the Patrick Clements clinic of the Central Middlesex Hospital (CMH) which was the local Venereology clinic in Acton, west London. The patient population there was very different from that at UCLH, there being a large West Indian community in the locality. The consultant at CMH was Joan Nabarro, who I considered a relic of British colonial Empire days. Joan enunciated English as the Queen did in the 1950s. Thus 'often' was 'orften', 'happy' was 'heppy' and 'my' was 'meh' to choose just a few examples. She was dictatorial and never quite had a handle on her subject. That said, I believe she was very hardworking and often looked tired. On one occasion she fell asleep while reading microscope slides and singed her hair as her head dropped gently into a Bunsen burner flame.

Driving from UCLH to CHM was not too difficult even though my red Austin 'Mini' had seen better days. The Marylebone Road and Westway connected the two hospitals. Parking was free for medical staff at both sites.

We had our fourth child in 1978 – our first girl – we named her Chani. Because of my depressive illness we felt we would not be able to cope with more children, so Tricia had her fallopian tubes tied where Chani was born – at the Royal Free Hospital. This did not make Tricia happy, in fact quite the reverse. She became depressed. She was put onto antidepressants and lorazepam. However the depression was about not being able to conceive, not an endogenous mood issue, so the medication could only do so much. I arranged for her to see Dr Heinz Wolff, the senior psychiatrist at UCLH, who was a very pleasant man, full of wisdom and a great listener and a support for Tricia. He had emigrated to the UK just before the War, originally studying mathematics at Cambridge and only later on switching to medicine. Although he had little formal psychiatric training, he was universally accepted as a doctor who had immense natural understanding of the human struggle and great compassion. After many months of carefully assessing Tricia he agreed that she should have a tubal reversal, but, he went on, she needed to be in a better state of health. So she came off her various psychotropic medication, including the lorazepam, which was a protracted nine month struggle, and obtained an appointment with Professor Robert Winston at the Hammersmith Hospital Fertility clinic. He thoughtfully gazed into the air – then expressed his view that the reversal might not help the depression, but agreed he would undertake the operation nonetheless. A year after the reversal our fifth child Leah-Miriam was born in 1982.

Reports of hitherto healthy young people suddenly becoming ill from rare infections such as pneumocystis carinii pneumonia or unusual skin diseases such as Kaposi's sarcoma were being whispered among the staff and patients of Venereology clinics in the early 1980s. They were finally documented in the major medical journals from 1981 onwards. These outbreaks were initially centred in San Francisco and New York amongst the male gay community, as well as some haemophiliacs and intravenous drug users. By early 1982, only a handful of similar patients had presented in the UK. There was a great deal of uncertainty about the cause of this illness in the early 1980s, but what was clear was that the patients were 'immunocompromised', in that their immune system, in particular their CD4 lymphocytes, were significantly lowered. Hence it was named the acquired immune deficiency syndrome (AIDS). It took a few years till the culprit – the human immunodeficiency virus – HIV – was isolated. The causative virus was eventually traced back to the Democratic Republic of Congo, where a man had become infected in 1959. However it is possible that humans were infected as early as 1910.

A long and bitter dispute between the Frenchman Luc Montagnier and the American Robert Gallo festered for many years, both claiming to be the first to isolate HIV in the lab. A Nobel Prize for Montagnier and a scientific censure placed on Gallo, seemed to resolve this issue. Gallo later accepted that Montagnier was the first to isolate HIV, and Montagnier accepted that Gallo had been the first to make the connection between AIDS and HIV.

HIV is a retrovirus, meaning that the virus converts a copy of its RNA genome into DNA, which is then incorporated into the DNA of its host. The prefix 'retro' refers to the backward conversion of RNA to DNA, whereas in the case of most viruses, their DNA is converted into RNA. Back in the early 1980s many ideas were put forward to explain the causation of AIDS. As a majority of the cases in the USA were gay men, attention was focused on why that group might be disproportionately affected. Amyl and butyl nitrite usage, and infection with cytomegalovirus were cited as possible causes. At that time the gay men I was seeing in clinic in London were physically well but very worried.

Bryn, a tall, muscular 27-year-old gay man who spoke with a lilting Ballymena accent, saw me in clinic in the summer of 1982. He spoke slowly and deliberately, picking nouns and adjectives with precision to describe his story. He was completing a PhD in medieval English at Oxbridge and was obviously as bright as a button, but also highly anxious about the reports emerging from the USA. He has been on holiday in New York a year before and had 'done the scene' in a big way, with a lot of 'popper' use – that is amyl or butyl nitrite. The original chemical in 'poppers' was amyl nitrite, which is very similar chemically to its butyl relative. In the past nitrites were contained in glass mesh bottles which needed to be crushed to release the chemical vapour. The crushing resulted in a popping sound – hence 'poppers'. As well as resulting in a 'high', these substances relax musculature and hence facilitate penetrative anal sex. They have a medical use in the form of amyl nitrite in patients who suffer from angina. They do not dilate cardiac blood vessel as originally thought as these blood vessels are not responsive in patients with heart disease. Their likely mode of action is by dilating veins in the body's periphery, thus causing pooling of blood leading to a decreased workload for the heart. But back to Bryn.

Bryn was now worried about his glands being swollen, and had 'funny spots' on the side of his tongue. I could feel no glands but he had pearly white plaques on both sides of his tongue. As our understanding of AIDS and HIV improved these tongue lesions were named hairy oral leukoplakia, and were found to be caused by Epstein–Barr virus (glandular fever virus). I reassured him that he looked well but there was an ominous non-verbal pessimism and fear in our exchanged glances. He had read that cytomegalovirus (CMV) was a possible cause of AIDS, and therefore asked me to send off a urine test for that – it came back negative. He died of AIDS two years later.

When early reports of AIDS appeared, I was still working at UCLH. On the assumption that CMV was in some way causative, I and colleagues in the haematology department looked at blood antibody levels of CMV in gay men (46 subjects) as well as controls who were heterosexual (76 subjects). We found a higher rate of significant levels in the gay men (76 per cent versus 50 per cent respectively). In no case could we grow CMV from urine. Around 30 per cent of the gay men used poppers regularly. Total white cell counts were no different in the two groups – we did not have access to CD4 white cell counts at that point. I concluded in the article – rather rashly, that there was no evidence that the 'immunocompromise syndrome' was present in London in 1982. In reality of course it was already there – an epidemic waiting to break!

Zhu, T, Korber, BT, Nahmias, AJ et al. An African HIV-1 sequence from 1959 and implications for the origin of the epidemic. Nature. 1998 Feb 5;391(6667):594–7.

Goldmeier, D, Linch, D and Mellars, BJ. Immunocompromise syndrome in homosexual men. Prevalence of possible risk factors and screening for the prodrome using an accurate white cell count. British Journal of Venereal Diseases 1983 Apr;59(2):127–30.

Jaffe, HW, Bregman, DJ and Selik, RM. Acquired immune deficiency syndrome in the United States: the first 1,000 cases. Journal of Infectious Diseases 1983 Aug;148(2):339–45.

Gallo, RC and Montagnier, L. The discovery of HIV as the cause of AIDS. The New England Journal of Medicine 2003 Dec 11;349(24):2283–5.

As I got used to my new jobs at UCLH and the Central Middlesex Hospital, my mother became increasingly ill. Her dentist had gently chided her for not keeping her tongue still while he treated her. She later developed weakness in her arms and legs. This worried me a great deal. I asked a UCLH neurologist, Dr Gerald Stern to see her. He was an internationally renowned expert on Parkinson's disease and was a pleasant and compassionate physician who had risen from humble beginnings in London's Jewish East End. He confirmed she had motor neurone disease (MND). Stephen Hawking's MND had only just started at that time. He was an unknown post doc researcher in the late 1970s. In the USA MND used to have the eponymous title of Lou Gehrig's disease, named after the famous baseball player who developed it in his late 30s and died within two years. The motor nerves in the brain and spinal cord are affected so that the disease produces increasing weakness but without sensory loss. The sufferer's mind is kept crystal clear, which in many ways is an added burden. There is still no treatment. Eating and drinking was a torture for my mother, as the muscles co-ordinating swallowing are affected, leading to choking and inhalation of food and drink. She lost weight both because her muscles shrivelled and because she now had a poor nutrient intake. I was devastated by her diagnosis, but had to remain cheerful and optimistic around her. As per her usual determined personality, she soldiered on in stoical fashion, being more concerned with the welfare of her children and grandchildren than her own health. She gradually became weaker. I therefore visited her every night on my way home and inserted a nasogastric tube in order to enable liquid feeding without the need for her swallowing. She told me, 'I am not afraid of dying, I am afraid of the Nazis'. Her speech later became incomprehensible. Her inability to make herself understood resulted in anger and frustration. Her last few days were spent on the wards at UCLH where she was given opiates to lessen her anxiety. She died in February 1979 and was buried in the United Synagogue cemetery in Bushey. As is the Jewish custom in the house of mourning my brother, father and myself sat 'shiva' for a week at my parents' home in Pennine Drive. This meant us sitting on low chairs for most of the day for a week. Friends and family came to console us. The orthodox Jewish way of doing this is to wait until the bereaved initiate conversation, as was described in the Book of Job. Job lost all his possessions and family. The three visitors that came to him sat in silence for a week until Job spoke. Jews pray three times a day in a quorum of ten men. These services are led, where possible, by the mourners during the week of 'shiva'. Again I found this very stressful and stumbled over the prayers. And yet I owed it to my mother to do this and lead the prayers. I did not miss any of the services for the 11 months of mourning. Tricia, who until this time had been agnostic, was inspired by the events around the shiva and how we kept to the letter of the Jewish law. She told me she wanted to learn more and be more active in keeping the religion. I advised her where to go. As her training analysis became less and less meaningful, so her religiosity and interest in Judaism increased.

After four years as senior registrar at UCLH, a vacancy came up as consultant colleague to Dr Oriel. I applied for this post, thinking that, as the incumbent senior registrar I would likely get the job. However there was competition in the form of Dr Tony Boakes, who had at one point been an anaesthetist. In the event he was successful, largely, I found out because Dr Ridgway, the microbiologist, was fearful that I 'would turn the clinic into a sexual dysfunction service'. Maybe in part that was true! David Oriel seemed almost embarrassed when he spoke to me later, saying that I would eventually find a job somewhere.

My interactions with David Oriel after this failure seemed closer – more like a father and son. He was always a very private person. Notwithstanding that, he not infrequently invited Tricia and myself to his flat in the Finchley Road, where I met his rather more outgoing wife Joan. He told me he enjoyed hill walking and classical music, particularly Wagner. David Oriel retired in 1988 after which he embarked on writing his magnum opus, *The Scars of Venus*, which is an illustrated history of venereal diseases. We not infrequently discussed the history of infections, in particular of gonorrhoea, after I had managed the complex case as delineate below. He informed me that James Boswell, the 18th-century Scottish diarist, lawyer and biographer of his friend Samuel Johnson, had died of kidney failure secondary to a urethral stricture likely subsequent to chronic gonorrhoea. A recent article suggests that modern gonococci originated in Europe or Africa in the 16th century and were subsequently disseminated globally, eventually becoming almost antibiotic resistant in the 21st century.

Sánchez-Busó, L, Golparian, D, et al. The impact of antimicrobials on gonococcal evolution. Nature Microbiology 2019 Nov;4(11):1941–1950.

GONOCOCCAL SEPTICAEMIA

In 1981, while working at UCLH, I was asked to see a patient at the Hospital for Tropical Diseases (HTD) in London, which at that time was housed in Pancras Road near what is now the new St Pancras International station. The hospital's origins lay in a meeting of the Seamen's Hospital Society on 8 March 1821, at which it was decided to establish a floating hospital for injured seamen, in particular merchant seamen and those in the fishing industry, as well as their dependants. The Admiralty allocated the Society a de-masted former naval ship, described as a 'hulk', in the form of HMS Grampus, which had formerly been a 50 gunned warship. (The largest 'ships of the line', such as HMS Victory, housed over 100 guns.) After ten years the hospital accommodation on Grampus was deemed inadequate and so in 1831 patients were transferred to the larger hulk HMS Dreadnought (not to be confused with the Dreadnought class battleships built just prior to World War 1). In 1870 the patients were transferred ashore to the land based infirmary in Greenwich which was named the Dreadnought Seamen's Hospital. In 1920 the Dreadnought Seamen's Society provided funds for the establishment of the Hospital for Tropical Diseases, which eventually moved to St Pancras. A section of the hospital in Greenwich, still named the Dreadnought Seamans Hospital was in the past devoted to STDs.

In 1975 I was asked to do a locum session at that very STD clinic. I was surprised that even then, a few years before it closed, it was run on military lines. It only saw male patients who were marched by two nurses into a large room, where I took their history in the presence of the nurses. When I asked them to drop their trousers so I could inspect and examine their genitals the request was repeated and barked out by one of the nurses.

To revert to 1980 and the HTD in St Pancras. The patient, a previously fit 32-year-old ruddy faced merchant seaman, sat contemplating his spatulate fingers, thick blue woollen beanie almost covering his eyes. I introduced myself. His lifted his gaze for a moment, grunted, then resumed his laconic digit observation. He was admitted to the wards of HTD, having recently returned from Lesotho complaining of a febrile illness, with headaches and pains in his wrists, hands and right shoulder. His hands were swollen, with pain on moving. He had an extra heart sound – a murmur, and a raised erythrocyte sedimentation rate which suggested he likely had an infective or inflammatory process going on. A large battery of tests for infections proved negative. By day 16 of his admission he had developed an abscess over his right biceps muscle. This was drained under anaesthesia. I was asked to get involved at this stage as gonorrhoea was found on culture. The gonococcus was also later found in his throat and urethra, even though those sites appeared normal. Before starting penicillin he had developed a painful swollen ankle. After an extended course of penicillin all his symptoms and abnormal signs resolved.

He had likely acquired the infection from sexual partners in Lesotho.

Gonococcal septicaemia presenting as an abscess is rare, occurring in under 1 per cent of cases of gonorrhoea. My patient ended up in HTD as he had a fever and had recently arrived from southern Africa. If he had complained of genital symptoms he would have initially gone to an STD clinic. It is likely that the abscess spread from the shoulder joint into surrounding muscle. Gonococcal septicaemia is often associated with deficiencies of complement – a sugary protein made by the liver, which helps to combat infections. Nowadays the term 'sepsis' has replaced the word septicaemia. It should be treated vigorously as soon as the diagnosis is established, as it can become life threatening.

Owino, NO, Goldmeier, D and Wall, RA. Gonococcal septicaemia presenting as a subcutaneous abscess. British Journal of Venereal Diseases 1981 Apr;57(2):143–4.

Masi, AT and Eisenstein, BI. Disseminated gonococcal infection (DGI) and gonococcal arthritis (GCA): II. Clinical manifestations, diagnosis, complications, treatment, and prevention. Seminars in Arthritis and Rheumatism 1981 Feb;10(3):173–97.

In spring 1982 the annual meeting of the Medical Society for the Study of Venereal Diseases (MSSVD), the national umbrella organisation for venereologists at the time, took place in Lubeck in West Germany. Lubeck is famous for its marzipan, which was offered freely in many parts of the conference centre and in our bedrooms. Apart from the sessions, during which I presented a paper on penicillin levels in the cerebrospinal fluid, we paid a trip to the extended wall between East and West Germany that started in Lubeck and extended southwards along the whole border. Dr Willie Harris, who was a consultant at the Venereology clinic of St Mary's Hospital Paddington, came up to me and chatted afterwards. In spite of me feeling that my stammer had let me down, he encouraged me to apply for a consultant job at St Mary's, knowing and openly accepting my interests in psychological medicine and sexual dysfunction. I interviewed for that job later on that year. The interview panel included Willie Harris, Joan Nabarro and Professor Michael Adler who was a medical epidemiologist specialising in Venereology. He was the first person to occupy a high level academic position in the speciality. He went on to change the name of the speciality from Venereology to Genitourinary Medicine (GUM), feeling that the former term was pejorative and associated with clinics hidden in basements with poor facilities. In some ways he was correct. He edited a *British Medical Journal* series on GUM and asked me to write a section on psychological aspects of the speciality. I think I did reasonably well at interview and so got the job. Apparently Joan Nabarro said I looked like 'Moses receiving the Ten Commandments' at the interview. Apart from the fact I had a beard I was unsure what she meant and took it as an anti-Semitic sleight. Her husband, Sir John Nabarro, was born into a famous Spanish, Sephardic Jewish family but practised as a Christian. I had always felt that Joan Nabarro, his wife, tutted under her breath silently when I took off Jewish holidays as annual leave.

Chapter 9

EARLY CONSULTANT YEARS

So I started my life as a consultant at St Mary's Hospital Paddington. Truth be known, I did not feel quite up to the job, in spite of my wearing a three piece suit rather than a white coat. My knowledge and patient care was good, but communication in formal lectures was, as usual, wanting.

The Venereology unit at St Mary's had recently been transferred from a basement to a purpose-built one-storey structure erected on stilts above one of the hospital car parks. In later years charitable monies enabled the addition of a red bricked, three-storey research building next to, and connecting to the NHS unit. In 1988 the unit was transferred to a new, three-storey building next to Outpatients. It was named the Jefferiss Wing. In that same year St Mary's Hospital Medical School was incorporated into Imperial College, along with Charing Cross and The Westminster Hospital Medical schools. The College was now known as Imperial College of Science, Technology and Medicine.

Dr Oriel had told me when I applied for the job that it was likely the busiest unit in the UK in the early 1980s! And it was. There were separate male and female waiting room, with up 50 people in each at any one time. No appointments were needed. The unit was called the Praed Street Clinic, the hospital itself being in Praed Street which was named after the chairman of the company that built the extension of the Grand Union Canal just to the north of the hospital in the 19th century. I had my own consulting room, and spent a great deal of the time either doing clinics or giving advice to junior doctors or nurses. If I was the consultant supervising the evening clinics, I stayed on till the last patient and last member of staff had exited the building. The cleaners arrived later on, to clear the discarded food, drink and other patient detritus that had accumulated during the day.

A VIOLIN CASE

It had been the end of a very long day. I had seen my last patient. As I wearily walked towards the exit I noticed a dark object beneath the metalled waiting room benches. I pulled out what appeared to be a violin case. Its outer chestnut brown skin was peeling. Preserved sections were heavily plastered with a myriad of stickers of various shapes and designs – scenes of places visited and cameo art. The hinges squeaked as I cautiously opened it. The case's dark red velvet lining had faded to grey in all but a few areas. The ghost of a violin lay among the case's litter. One of the strings had snapped, its loose ends lying serpentine in disarray. The chinrest's edges were jagged. The bridge lay at an angle to the upright. The once wooden tail piece had been replaced with a metal substitute. I gently lifted the violin, beneath which was a frayed bowstring atop numerous sheets of paper and a battered cloth cap containing a few coins. I separated the various contents of the case. My attention alighted on sheet music for Brahm's violin concerto. Searching the contents further I found a bottle of pills labelled 'olanzapine' and an appointment card for a psychiatric consultation for the prior week but without a patient name. But I knew who the owner was. Pinchos, a middle-aged fiddle player who had once upon a time played in the second strings of a well-known national orchestra. His paranoia and subsequent diagnosis of schizophrenia had ended his professional working days. He came to my unit at St Mary's to play merry tunes to a captive waiting room. Between each solo he would regale the audience fortissimo with vivid and detailed accounts of his most recent sexual escapades. Some listeners were entertained, others just embarrassed. We often had to gently escort him off the premises after checking that he had not attended for a medical reason. His reply was always, 'No just for company'. My charge nurse colleague knew his full name and address, and returned the case and contents to him.

Because of the topic of my thesis, nurses and junior doctors would ask me for my opinion on men with proctitis. Some patients with proctitis were more memorable than others.

I first met Patrick in 1982 when I was a young consultant at St Mary's Hospital. He spoke with a lopsided smile and carefully chosen words, and was always clean and smartly dressed with a penchant for colourful waistcoats, plucked eyebrows, peroxide blond hair and well placed eye make-up. Patrick would often bring his latest boyfriend with him, almost as an offering to me in terms of contact tracing. He was sexually very adventurous, pronouncing that as he had given up drugs and alcohol he felt he had to splash out somewhere. All this was just prior to the outbreak of HIV in the male gay community. One adventure with a partner involved connecting up a flexible hollow tube anally between himself and his current beau, but not before placing his pet gerbil into the tube. Seemingly he found the animal's foray into his rectum very pleasant. Unfortunately the poor creature could not find its way out and succumbed. At that point the fun stopped. Natural bowel evacuation failed to dislodge the body after which he came to see me for help.

My Surgical colleagues admitted him as he needed a general anaesthetic to enable per anal instrumentation to remove said body (came out in sections). A few months later a vibrator disappeared in similar fashion. Fortunately I could see the handle through a proctoscope and was able to retrieve it. After all this trauma to the anorectal area I felt he needed to be followed up regularly. Some months after the vibrator episode he still had mild inflammation of the rectum. I sent off quite a battery of investigations to exclude infections. The only one to come back as positive was for Entamoeba histolytica. I treated him for this. However the mild redness of the rectum persisted – I felt likely secondary to trauma. Patrick unfortunately became infected with HIV, and died about two years later, this being before successful treatment became available. We lost many colourful characters in those early days of HIV.

As a new consultant I was keen to carry on research into the possible causes of rectal inflammation in gay men. And I had met Entamoeba histolytica (EH) before. EH is an intestinal parasite that passes from person to person via the faeco-oral route, typically by eating or drinking something contaminated. It is estimated to infect up to 50 million people worldwide at any time and is commonly found in the tropics. EH can cause serious inflammation of the large bowel including the rectum as well as the liver and many other organs. It was one of the many bowel diseases seemingly causing symptoms in gay men in the 1970s which were at that time grouped together using the epithet 'the gay bowel syndrome'. Looked at from the 21st century the diseases they described were neither gay specific nor was it a syndrome. The term is now considered derogatory. Nevertheless its reporting did focus clinicians' minds on infections that could be passed on by oro-genital or oro-anal sex. In my last year at UCLH David Oriel and I had been approached by Peter Sargeaunt, a senior technician at the London School of Tropical Medicine and Hygiene, to provide him with stool samples from gay men. He was of the opinion that Entamoebae histolyticas (EH) were of differing subtypes or zymodemes, some causing illness and others just bowel passengers that caused no problems. Faecal samples from gay men at UCLH were compared to samples of sick patients in Africa, Asia and South America. All the samples from the latter three continents were pathogenic. Those from the gay men were not. After I settled in at St Mary's I obtained pharmaceutical company research grants and ethics committee approval to undertake further research into EH looking for evidence that EH in gay men caused inflammation via rectal invasion. I teamed up again with Peter Sargeaunt. Peter had originally worked in the laboratory of the Dreadnought Seamen's Hospital in Greenwich, which was near the London Docks where seamen who returned from the tropics with dysentery were provided with diagnosis and treatment. Another of Peter's talents was woodwork. He built a 10-m boat in his back garden which on completion had to be hoisted by crane over his house!

To return to our study – we defined invasion of the colon using four criteria – finding EH and inflammatory changes in biopsies of the rectum, looking for blood inside the parasite on microscopy, high antibody levels to the protozoan in blood and looking for the enzyme type that is usually associated with disease. Peter undertook the enzyme type investigation in his lab. No blood within the parasite or significant serum antibody was found in 23 EH excreters of EH or 11 controls. There was no significant difference in the histological appearance of the rectal tissue comparing EH secretors to controls. The enzyme types in all cases were non pathogenic. That being so we concluded that the majority of men having sex with men who excrete EH are in fact infected not with EH but with a non pathogenic variant now called Entamoeba Dispar. The paper was published in *The Lancet*. Similar work undertaken by my colleague Erica Allason-Jones at

University College London confirmed our findings.

Of course it is still possible to be infected with the pathogenic strain. If the protozoan is seen in rectal material nowadays a DNA test would confirm or refute its pathogenicity.

Goldmeier, D, Sargeaunt, PG and Price, AB et al. Is Entamoeba histolytica in homosexual men a pathogen? The Lancet. 1986 Mar 22;1(8482):641–4.
Allason-Jones E, Mindel, A, Sargeaunt, P and Williams, P. Entamoeba histolytica as a commensal intestinal parasite in homosexual men. The New England Journal of Medicine. 1986 Aug 7;315(6):353–6.
Sargeaunt, PG, Oates, JK, MacLennan, I, Oriel, JD and Goldmeier D. Entamoeba histolytica in male homosexuals. British Journal Venereal Disease. 1983 Jun;59(3):193–5. doi: 10.1136/sti.59.3.193. PMID: 6303491; PMCID: PMC1046175.
Kazal, HL, Sohn, N, Carrasco, JI, Robilotti, JG and Delaney, WE. The gay bowel syndrome: clinico-pathologic correlation in 260 cases. Annals of Clinical Laboratory Science. 1976 Mar–Apr;6(2):184–92. PMID: 946385.

During the early 1980s AIDS became an increasingly common problem in the male gay community. Our unit at St Mary's de facto became a national centre for the management of patients with AIDS and after a diagnostic blood test became available in 1985 for anyone who tested positive for HIV. The physicians who were actively involved in their management were Willie Harris, Tony Pinching an immunologist and Jonathan Weber, who was a junior doctor at the time but had the foresight to gather the patients together into a research cohort that was eventually funded by the Wellcome Trust. Until the advent of the first anti-HIV medication AZT in 1987, treatment of patients with AIDs consisted of management of the infections or diseases that would eventually cause their demise. A common presentation was with pneumocystis carinii pneumonia, abbreviated to PCP. PCP is caused by a fungus that attacks the lungs of people whose immune systems are compromised. Kaposi's sarcoma of the skin and internal organs was treated by X-ray and chemotherapy. As patients became sicker they ended up in the inpatient wards – either Almroth Wright and Rodney Porter. Between 1982 and 1985 a fifth of all the inpatients at St Mary's had AIDS, with at least one death occurring a day. A word about the men after whom the wards were named. Almroth Wright (1861–1947) worked at St Mary's Hospital. He was knighted for research in the fields of bacteriology and immunology, and developing a successful typhoid vaccine. He was a contemporary and friend of a fellow Irishman, George Bernard Shaw. Shaw credits Wright as the source of the medical information he used in his play 'Doctor's Dilemma' where Wright was immortalised and gently caricatured as Sir Colenso Ridgeon. The play makes reference to the immunological, pre antibiotic theories of the day saying 'stimulate the phagocytes, drugs are a delusion'. Wright was a scientific giant who nevertheless had his personal foibles, including his opposition to women's suffrage. Behind his back he was referred to as 'Sir Almost Right'. Rodney Porter was a research immunologist at St Mary's Hospital Medical School who won the Nobel Prize for medicine for elucidating the structure of antibodies.

The doctors and nurses who treated patients with AIDs in those early pre anti-HIV therapy days needed a strong psychological constitution, bravery, up to the minute knowledge and compassion.

I was particularly interested in the neurological and psychiatric manifestations of AIDS.

PSYCHIATRIC ILLNESS AND HIV

Colin was a 20-year-old gay man who had been HIV positive for two years. As he spoke to me in his broad cockney accent, he would intermittently blow his long blond hair out of his field of vision, using his lopsided mouth as an air puffer. His slim and small stature, with teenage looks, obviously appealed to many suitors. In 1989 the only drug available to treat the condition was zidovudine, also known as AZT. As he became sicker he was changed to DDI, a similar medication. Colin first had sex with a man when he was aged 15 and from then on enjoyed a varied sex life with both regular and casual partners. No holds were barred in terms of what he did at sex. He was very open about his sexual tastes and described his sex life to his mother, who had brought him up on her own since her husband had left the family home when Colin was three. A year before I met him he could no longer work in a greengrocers shop because, from time to time, he forgot what customers had asked him from one moment to the next. By the time I saw him he had developed full-blown AIDS-related dementia, as it was called then, and had almost no short term memory. However he still knew his mother. He came to clinic three times a week for respite care. I asked his mother how he had been doing at home. 'He is generally quiet but when he becomes agitated I know it is time to "do him"'. Being unclear what she meant by this I asked further. 'I wank him – he sleeps like a baby after that'. He died two years after contracting HIV.

Pierre was my patient from mid 1988 onwards. He was a 40-year-old French Canadian, who had been monogamous with his boyfriend for 15 years, and was a talented graphic designer, who was invariably dressed smartly in dark suits, which matched his short cropped black hair and Van Dyke beard. His speech was accurate and delivered slowly and to the point, with a tinge of a French accent. He did not question me about how he might have acquired HIV but knew his partner travelled outside the UK a lot. He initially attended clinic because of genital warts. A routine HIV test confirmed he had the infection. After hearing the diagnosis he felt hopeless for a few weeks, had episodes of crying and felt nervous at times. His work suffered. His clinical picture was likely due to an adjustment disorder. After two months these symptoms subsided and he continued to lead a happy and productive life until he developed full-blown AIDS in 1990. He died a year later.

Because of my interest in psychiatry, I encouraged the development of links between the GU Medicine department at St Mary's Hospital and the psychiatric unit, in particular in relation to patients with HIV infections. Psychiatrists largely treat mentally sick patients with medication, as opposed to psychologists who use talking therapies of one sort or another. Dr Kenneth Granville-Grossman was the consultant psychiatrist interested in HIV. At the time he worked with me he was in his mid sixties. He assessed every patient under his care with slow, yet accurate deliberation. A few years later he retired and became a keen aviator having obtained his pilot's licence while still at work. The psychiatrists and I looked at all HIV cases referred to their unit. We described 60 such patients seen over the year ending July 1989. This compared to about 100 HIV patients seen in the same time by the psychology unit (mainly for anxiety and adjustment disorders). The 60 patients constituted just under 2 per cent of the total who were HIV positive and 10 per cent of those with AIDS. Depressive illnesses were the most frequently encountered problem – ten expressed suicidal intention. All received standard antidepressant medication. Three patients had hypomania (overactivity and excited behaviour that may endanger a patient's life) all of whom died within the following nine months, with evidence of brain atrophy in all on CT scanning. Those patients with full-blown AIDS often had more than one HIV related illness. Brain CT abnormality was also found in patients with AIDS related dementia and personality changes. There were three patients with personality disorders and associated aggression that was out of keeping compared to the patient's pre morbid personality. They were considered a risk to themselves when aggressive and were therefore admitted to the inpatient psychiatric unit for medication and safeguarding. Of the two schizophrenic patients in the group, one rapidly developed AIDS after becoming HIV infected. Nine patients were deemed to have affective symptoms (a depressed or manic mood) caused by an adjustment disorder i.e. a stress response to their physical illness). Not surprisingly, the risk of suicide amongst those with affective symptoms was nearly 40 times greater than that of the general population.

At the time we published this paper there was no effective long-term treatment for HIV or AIDS. AZT and later DDI could buy an extra year of life, but after a few months the HIV strain invariably became resistant. Not until the advent of using combination HIV treatments in 1995 were HIV sufferers able to live an almost normal span of years.

Seth, R, Granville-Grossman, K, Goldmeier, D and Lynch S. Psychiatric illnesses in patients with HIV infection and AIDS referred to the liaison psychiatrist. British Journal of Psychiatry. 1991 Sep;159:347–50.

With the advent of this new and lethal infection, it was not surprising that some anxious and ruminatory patients developed health anxiety in relation to AIDS. Health anxiety, previously known as hypochondriasis, is a not uncommon presentation to STI clinics. It can relate to HIV or indeed any STI, and is not easy to manage as the patient tends to be reassured transiently but soon relapses into fear of disease. Patients with health anxiety often go from one clinic to the next to obtain the short-lived reassurance that makes them feel better momentarily.

HEALTH ANXIETY, HYPOCHONDRIASIS AND SEXUAL INFECTIONS

The man's head hung between his knees as he blindly scratched and picked away at his paint stained coarse hands. His head shot to attention as I called his clinic number, revealing large, lop-sided, black rimmed glasses, a heavily furrowed face and a sweat drenched forehead. He stood up to his full height of about five foot six and pulled up his paint stained jeans, which would rise no further than the base of his abdominal obesity. Peter, a 65-year-old decorator, was keen on cricket. He particularly loved going to Lord's cricket ground to see England play. It was the summer of 1984 and the West Indies had come to visit. Ian Botham clocked up 4,000 runs that year. Midway in England's first innings my patient felt he had to open his bowels. Some water from the toilet bowl splashed up onto his anus. At that instant he became convinced he had caught AIDS via contaminated water. When I saw him two weeks after the event he had lost some weight, couldn't sleep, felt hot and at times sweated profusely. It was indeed hot that summer. He had no other risk factors for HIV, being happily married (to a woman) and not having had extramarital sex. He denied receiving blood products in the past. Examination showed him to be very anxious, but no signs of AIDS. There were no blood tests for HIV till the following year. He was certainly not at high risk of AIDS. I did not feel he was delusional rather he had health anxiety with an overvalued idea that he had acquired HIV. Reassurance helped temporarily. After many months he failed to reattend, and likely received a negative HIV antibody test in 1986.

Imogen flexed her torso from her supine position on the examination couch. I sensed fear and anger as her large brown eyes shot daggers at me. She pulled at her right vulva with both hands, her long black hair sweeping over her face and focused on a what looked to me like normal skin genital skin, with possibly an area of mild healed scarring. Her voice, now a shout, was uttered an octave higher than the dulcet tones in which she gave her history.

'You are not looking properly doctor. They are warts. Can't you see?'

Imogen was a 25-year-old postgraduate student who had been initially diagnosed with genital warts. After they had been treated successfully she returned repeatedly because she believed some of the small scars left at the site of the warts were in fact a new warts. In spite of being reassured at five separate clinic attendances that the scar did not represent an active wart, she continued to insist to the contrary. Reassurance each time was transient, lasting an hour or two. She later divulged that she had been to two other STD clinics to try to persuade the clinicians there to treat the scar as a wart. She declined psychological help and was lost to follow up after three months.

In 1983 Jonathan Weber (who became the professor of GU Medicine and HIV and is now Dean of the Medical School at Imperial College) and I had a short article published in the British Medical Journal discussing our experience of patients presenting with 'Fear of AIDS' rather than actual AIDS which at that time was rare. All three patients we reported on had no previous psychiatric illness but had read articles in the press or viewed television programmes about AIDS and then went on to develop anxiety, depression and ruminations, fearing that they had acquired the infection. Many of the symptoms of anxiety, such as sweating and weight loss, overlapped with those of AIDS. However, none were at high risk of developing HIV, although, again, at that time there were no blood tests that could be undertaken to exclude HIV. All three patients expressed suicidal ideas. They all improved on antidepressant treatment and were strongly advised to avoid speculative journalism.

Although actual physical illness may itself promote anxiety, the term health anxiety is usually taken to refer to patients who attend and then reattend clinics – in this case STI clinics – with ongoing concerns, despite an otherwise normal clinical picture and negative investigations and despite repeated ineffective reassurance. Health anxiety can be a feature of an obsessional illness, as well as being a prominent feature in patients with major depression. Delusional states can also present with health anxiety. However the general meaning of the term health anxiety is one where the fear of the disease is the primary issue.

In 2004 colleagues from psychology, psychiatry and I looked at the prevalence and service implications of health anxiety in STD clinics. The Health Anxiety Inventory (HAI) was used to screen 694 clinic patients. Significant health anxiety was found overall in about 10 per cent of the patients, with higher levels being found in those who did not have an STI. HAI scores were stable over time implying that the condition was chronic for many patients. Higher HAI scores were also correlated with more frequent clinic doctor appointments. Cognitive behaviour therapies and judicious use of antidepressants can successfully treat health anxiety.

Weber, J and Goldmeier, D. Medicine and the Media. British Medical Journal 1983;287:420.

Seivewright, H, Salkovskis, P, Green, J et al. Prevalence and service implications of health anxiety in genitourinary medicine clinics. International Journal of STD & AIDS. 2004 Aug;15(8):519–22.

Genital herpes continued to be an interest of mine at St Mary's. I set up a genital herpes clinic in 1982. The anti-herpes medication acyclovir had just received a treatment licence about that time. Acyclovir had been initially produced in 1974 and patented by Gertrude Elion on behalf of Burroughs Wellcome in the USA. She went on to win the Nobel Prize in Medicine. Acyclovir is preferentially taken up by human cells infected by herpes simplex where it terminates the viral DNA from replicating. It is remarkably safe when taken orally but if undiluted and given intravenously as a once-off shot it can crystallise out in the kidneys. Trials of the product's use in oral form continued at St Mary's for a number of years. It significantly shortens the length of first outbreaks and recurrences of genital herpes when taken as a five day course. When taken continuously it markedly reduces symptomatic recurrence rates and is safe. Taken daily it makes it much less likely that the infection can be passed onto a sexual partner. Its use certainly transformed the lives of many patients who had been celibate for years. In spite of acyclovir, many patients who came to this clinic remained psychologically devastated by their infection, feeling they were contaminated and that no one would ever want to have a relationship with them, let alone have sexual intercourse. Some patients needed ongoing support, psychological help and medication for many years.

I called her clinic number. A young woman wearing a red patterned bandana cautiously stood up. Her loose fitting black trousers and matching top concealed a slim frame. She shuffled into my consulting room, slowly and carefully sitting down. Tresses of light brown wavy hair cascaded down the side of the headscarf. Her moist, reddened eyes burst into tears as I enquired if she was in pain. Margaret held the sides of her chair as her story unfolded. She was a 23-year-old postgraduate student of fine art who had been in a nine-month relationship with a man who she felt was very special. More specifically she complained of continuous knife-like throbbing pain in her genitals over the previous two days, as well as difficulty passing urine. Her partner had no symptoms, and had never knowingly had a genital problem. I found her to have numerous small painful ulcers over the anus and vulva, told her she likely had a first outbreak of genital herpes and gave her a five day course of acyclovir tablets. Swabs taken for herpes simplex virus detection came back positive three days later. A neater, but paler version of Margaret strode into my room a week later, her hair tied back in a bunch, wearing pink lipstick and a touch of mascara. Her furrowed forehead and knotted eyebrows foretold her searching questions. Although her genital symptoms had resolved and her vulva had almost healed, she was distressed for a number of reasons. In particular she wanted to know from whom she had acquired the infection, would she be able to completely eradicate it and would she need a Caesarean section when she became pregnant. Sensible questions from a patient who had read up about her condition since last seeing me.

Before I answered her specific questions I enquired if she had suffered from anxiety in the past, to which she replied that she had been treated for 'a nervous condition with antidepressants' when aged 18. I told her that up to 80 per cent of people with type 2 genital herpes (she had this type) have no symptoms, so that although her current partner may have been silently infected her, it may have been any of her previous partners. In terms of eradication, this was impossible, as the virus remains inert as a string of DNA rather than a whole virus near the spinal roots, and could reactivate at any time. Acyclovir, I went on, prevents DNA replication but cannot destroy inert DNA. However if she did suffer from frequent recurrences then longer courses of acyclovir would supress it for as long as she and I thought it necessary. If she did not become pregnant in the next six months a Caesarean section was very unlikely to be necessary.

I continued to see Margaret over the next four years. She developed three recurrences of genital herpes in the first six months of follow up, after which I put her onto daily acyclovir for a year. With the help of a relationship therapist, her initial distrust of her partner mellowed into a successful marriage. She completed her PhD fours year after I first saw her.

Whitley, RJ and Gnann, JW Jr. Acyclovir: a decade later. The New England Journal of Medicine. 1992 Sep 10;327(11):782–9.

A word about nomenclature. By the time I became a consultant at St Mary's, the speciality I practised was no longer termed venereology, rather Genitourinary Medicine, the term, as discussed above, having been originally coined by Michael Adler. True enough, the term venereology did have connotations of poorly lit hospital basements, where shame-faced patients would slink into the department, face covered by sunglasses and coat collars turned up. But was Genitourinary Medicine any better? Some critics felt it had connotations of male urinals. Furthermore, the illnesses managed were certainly not restricted to the genitals. And it made little acknowledgement that there are always, to a greater or lesser degree, psychosocial aspects of even the most straightforward infections. But the term GU Medicine stuck, and the speciality is named GU Medicine and HIV by the Royal College of Physicians. And what type of diseases are managed in GUM clinics? It is treatment of sexually transmitted diseases (STDs) or more recently sexually transmitted infections (STIs). In continental Europe dermatology and venereology are usually managed in the same departments, with venereal diseases being the junior partner. After the Venereal Diseases Act of Parliament in 1917 in the UK, venereology was separated from skin departments, mainly as the government of the day saw how important VD was to the county's general health. As time went on at St Mary's I spent more time treating sexual dysfunctions such as erectile dysfunction and low sexual desire. The practice of treating those dysfunctions can be termed sexual medicine. Nowadays it is generally understood to use a biopsychosocial approach to disease management. Professor Alan Riley, of whom more later, championed the termed sexual medicine. He held a chair in that speciality at the University of East Lancashire. One also hears the term 'Sexual Health' which usually refers to sexual infections, but can be used rather more loosely to mean STIs, contraception and sexual medicine. I hope I have not confused you!

Life as a consultant in the 1980s was never uninteresting. Junior doctors came and went, the senior house officers only working with us for six months at a time. We attracted doctors who were in training for gynaecology as well as those who wished to specialise in medicine. In later years two or three GP trainees worked for six months at a time in our unit. In the 1980s and 1990s GU Medicine was gradually taking on some aspects of what was hitherto deemed to be gynaecology. One such area was taking cervical cancer smears, looking for precancerous changes in the cells. This made a lot of sense, as the usual types of cervical cancer, the squamous cell variety, is in large part caused by oncogenic (meaning cancer inducing) subtypes of the human papilloma virus (HPV) which are sexually transmitted. Other factors leading to cervical cancer include smoking and an impaired immune system. Different subtypes of HPV cause benign genital warts. By the mid 1980s there were both consultants and junior colleagues at St Mary's interested in cervical pathology, which could be assessed more accurately by using a colposcope, which is essentially the same instrument I used to obtain magnified images of the rectum.

Precancerous cells in the cervix of the uterus may be detected by mass cervical screening, although it is likely that current human papilloma virus vaccination of teenagers will prevent both cervical precancer and cancer and is generally considered safe. When more than very early stage cervical precancer is detected, it is usual to treat it by some method of ablation or destruction to prevent progression to cancer. This may be by use of a medical laser, cold knife or LLETZ (large loop excision of the transformation zone). These treatments have had a huge effect of decreasing cervical cancer, although they may be associated with post-treatment bleeding, infection, as well as pre-term labour and low birth weight babies. Possible sexual side effects of these procedures are rarely discussed.

In July 2018 Kate Orson bravely reported in the Daily Mail on her sex life after treatment for cervical precancer, which had taken place 13 years previously. At that time she was found to have abnormal cells in her cervix, and was treated by LLETZ. LLETZ uses heat to cut away precancerous tissue in the cervix to a depth of 8 mm. After the procedure, Kate had lost all sexual desire and was unable to orgasm. She also suffered a number of other symptoms including pain on resuming sex, tension and exhaustion. My colleague in San Diego, Irwin Goldstein commented that LLETZ is not treating cancer but precancer. He felt that a watch and wait management would produce no side effects in the long run, and be just as effective in many women. Another woman who had undergone LLETZ went to see Irwin Goldstein and was found to have 'genital numbness' on physical examination.

It is important to consider anatomy in the possible explanations of these women's toxic responses to LLETZ. The cervix of the uterus, according to the Royal College of Obstetricians and Gynaecologists, is innervated by the inferior hypogastric plexus of nerves, which itself connects to the hypogastric nerve, pelvic plus of nerves and finally to the vagus nerve which connects to the brain. Barry Komisaruk, a researcher in the USA, has found that women who have had spinal cord injuries can still have orgasms via cervical stimulation. Brain confirmation of their orgasms was demonstrated in that 'sexual areas' such as the medial amygdala, fronto-parietal cerebral lobes and hypothalamus lit up on MRI scans as they orgasmed. That being the case, it is presumed there is a sexual spinal cord bypass via the cervix and the hypogastric gastric nerve. By removing part of the cervix at LLETZ it is possible that the associated hypogastric nerves are affected. This could explain the lack of arousal, inability to orgasm and even lack of lubrication at sex which could lead to pain. It is not so easy to explain the low sexual desire by direct involvement of the LLETZ. Radmila Sparic and her colleagues from Belgrade have recently followed up 146 women who either underwent LLETZ or similar excision by cold knife. On follow up 28 per cent became less interested in sex and also had higher rates of anxiety and depression. Those women in whom there were 'complications' after cervical excision had significantly higher rates of low sex drive, as well as anxiety and depression. The type of treatment did not affect outcome.

The exact mechanism of a bad sexual response to LLETZ is therefore unclear. Interruption of nerve supply and 'retrograde death' of the nerves nearer the brain may take place. The cause of low desire, anxiety and depression may well be secondary to impaired sexual response.

That said, I am keeping an open mind about LLETZ and its potential pathological sequelae.

Orson, K. The women who say treatment after a smear test ruined their sex lives. Daily Mail 17 July 2018.

Prendiville, W. The treatment of CIN: what are the risks? Cytopathology. 2009 Jun;20(3):145–53.

Sparić, R, Papoutsis, D, Kadija, S et al. Psychosexual outcomes in women of reproductive age at more than two-years from excisional cervical treatment – a cross-sectional study. Journal of Psychosomatic Obstetrics & Gynaecology. 2019 Jun;40(2):128–37.

Jørgensen, L, Gøtzsche, PC and Jefferson, T. Benefits and harms of the human papillomavirus (HPV) vaccines: systematic review with meta-analyses of trial data from clinical study reports. Systematic Reviews 2020 Feb 28;9(1):43.

Komisaruk, BR, Whipple, B, Crawford, A et al. Brain activation during vaginocervical self-stimulation and orgasm in women with complete spinal cord injury: fMRI evidence of mediation by the vagus nerves. Brain Research. 2004 Oct 22;1024(1-2):77–88.

Chapter 10
MORE ON SYPHILIS

Throughout my time at St Mary's I retained my interest in syphilis and was considered somewhat of an expert on the disease. I guess this was because I had worked for Eric Dunlop, had been involved peripherally in his syphilis research, and had seen his patients with end-stage disease who were used at teaching subjects at The London. In spite of this apparent expertise, there was a great deal about the infection that I did not know, and had to look up in current textbooks. After 1996, literature searches could be done via the online medical search engine PubMed. By the 1990s antibiotics had been around for 50 years, so that although the presentations of early disease were the same as those described in the pre-antibiotic literature, signs of late syphilis could be modified by intercurrent antibiotics used to treat other infections in patients who did not know they had been exposed to the syphilis earlier on. The textbooks current at the end of the 20th century almost always gave short accounts of the features of late syphilis, probably lifting them verbatim from other older texts. What I needed was a good reference book, describing the features of all stages of the infection as they appeared in the pre-antibiotics era. *Modern Clinical Syphilology*, also known as 'The Green Monster' came to the rescue.

The senior author of the book, John Hinchman Stokes, an American, was a highly intelligent, eagle-eyed doctor who documented what he saw, collaborating with his other USA colleagues to produce the excellent descriptive text. Stokes, who came from a Quaker family, was described as somewhat of a martinet, insisting, as his father had told him, that shiny shoes and clean shirt cuffs and collars were de rigueur for all professionals.

Modern Clinical Syphilology

The edition I possessed was published during World War 2. It was a large, green, buckram covered tome that lay on my desk at St Mary's at all times. It possessed a haunting odour of yesteryear, an acidic mustiness, with hints of vanilla and drying cut grass. Others have described old books as smelling of chocolate, coffee, or 'earthy' or 'woody'. Over the years I learned to associate the book's aroma with its content. Modern Clinical Syphilology *had been passed on to me by Dr Jimmy Jefferiss in 1990. By 'modern' it meant 1944 ! Dr Jefferiss had been the senior consultant at St Mary's STD department for many years. By the time I knew him he had retired but had set up a charitable trust to enable research work to be undertaken into HIV at St Mary's Hospital Medical School (later to become Imperial College) as well as at Liverpool University. He told me that while still at school he drove up from Kent in an open-top sports car to be interviewed for a place at St Mary's Hospital Medical School. I guess that must have been in the late 1920s, when he would almost certainly have been interviewed by the Dean, Dr Charles Wilson, who later became Winston Churchill's personal physician. Wilson was elevated to the peerage as Lord Moran, and was president of the Royal College of Physicians for many years. To revert to the textbook, it was obviously out of date in many ways, but was nevertheless a fount of clinical knowledge that became unavailable to current clinicians, blood screening and antibiotics having dramatically altered outcomes. It was also a history of syphilis in the first half of the 20th century.*

Syphilis had been recognised by its clinical signs for many centuries and named 'the great pox' (as opposed to the viral 'smallpox') although there was some confusion and overlap between it and yellow urethral discharge (gonorrhoea) before the 19th century. Early treatments had included prolonged mercury use in the form of ointments – hence the old adage 'Venus for a night – Mercury for the rest of your life'. Mercury had terrible side effects causing nerve damage, kidney failure, severe mouth ulcers and loss of teeth. Many patients died of mercurial poisoning rather than from syphilis. Nowadays one is most likely to suffer from mercury poisoning by ingesting too much seafood.

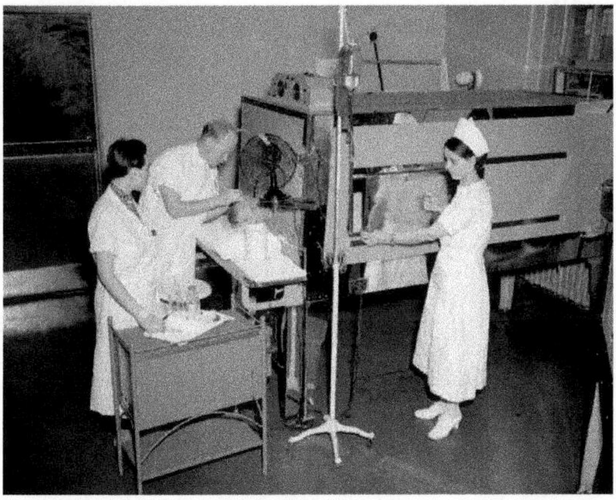

Heat Box

Professor Caesar Boeck in Norway stopped mercury being used as a treatment for syphilis in the patients he saw with early disease between 1890 and 1910. He believed that allowing the disease to take its natural course was better for patient outcome than mercury treatment. They were given no treatment even after the advent of Salvarsan, bringing into question the ethics of withholding even moderately successful therapy. Nevertheless he and subsequent doctors kept careful records of 2,000 patients who had early syphilis. This meant that over the ensuing years a natural history study of syphilis was built up and detailed, some of the patients being followed up for nearly 70 years.

Treatment in the earlier part of the 20th century had also included heating the person up to 105 °F- either in a heat box or by giving the patient a benign strain of malaria. After the malaria induced hyperthermia, the patient was treated with quinine to eradicate the malarial parasite.

The hyperthermia resulted in causative organism of syphilis treponema pallidum (TP) disintegrating – hopefully, without killing the patient! The gangster Al Capone was treated for syphilis in such a heat box while in prison.

TP was first identified in 1905 in Germany by Schaudinn and Hoffmann. A year later Auguste Wassermann, again in Germany, formulated a reasonably accurate blood test for syphilis based on the interaction of syphilis antigen (derived from the liver of a dead congenital syphilitic baby) with antibodies in the blood of the patient.

Unlike most bacteria, which can be grown easily in the lab, TP is very fastidious and until recently could not be kept alive (unless snap frozen). This was realised very early on, so that in 1906, inoculation of TP into rabbits' testes was successfully undertaken as described above, thereby enabling passage of the organism and facilitating scientific work on the organism.

In the late 1980s, Dr Philip Hay, who came to work in my department, used the new (at the time) technique of polymerase chain reaction (PCR) to detect the organism. PCR is very precise. DNA 'primers' are inserted onto single stranded TP DNA (the double strands are separated by heating) that mark the start and end of the target area. Then free nucleotides (the units that make up DNA) and a heat resistant enzyme (DNA polymerase) are added which attach to the target area of the single strand thus making double stranded DNA – one strand from the old section and one from the new. This is then heated to split it and the cycle repeated. PCR of TP thus produces millions of DNA copies. Phil Hay's work was one of the first to use PCR in syphilis research. In 2011 this work was carried forward in our unit by Dr Craig Tipple who used PCR to quantify the bacterial load or quantity of TPs that could be found in patients with syphilis. Craig's up-to-date PCR technique was able to detect a single DNA copy from syphilitic ulcers and blood of patients with secondary syphilis.

In February 2017 a group reported being able to keep TP alive in the lab for three years, growing them on rabbit epithelial cells in a special chemical mix and in a low oxygen environment. The TP grown under these very tight conditions were able to infect rabbits, thus showing they had lost none of their pathogenicity.

Stokes, JH, Beerman, H and Ingraham, N. Modern Clinical Syphilology. 3rd ed. Philadelphia: Saunders; 1944.

Hay, PE, Clarke, JR, Taylor-Robinson, D and Goldmeier, D. Detection of treponemal DNA in the CSF of patients with syphilis and HIV infection using the polymerase chain reaction. Genitourinary Medicine. 1990 Dec;66(6):428–32.

Tipple, C, Hanna, MO and Hill, S et al. Getting the measure of syphilis: qPCR to better understand early infection. Sexually Transmitted Infections. 2011 Oct;87(6):479–85.

Edmondson, DG and Norris, SJ. In Vitro Cultivation of the Syphilis Spirochete Treponema pallidum. Current Protocols. 2021 Feb;1(2):e44. doi: 10.1002/cpz1.44. PMID: 33599121; PMCID: PMC7986111.

Fyrand, O and Granholt, A. The history of venereology in Norway. Genitourinary Medicine. 1994 Jun;70(3):215–7.

After I retired, I left Modern Clinical Syphilology on the desk for my successor. It belongs to the department, not to me.

In 1985 I received a phone call from a radiologist at St Mary's asking for my opinion about radio-opaque shadows in the pelvic area of an abdominal X-ray of an 85-year-old man who had been admitted because of a confusional state. The radiologist had looked up the possible causes and found that one was past injections of bismuth used to treat syphilis. I agreed this was a possibility and I went to chat to the patient once the confusion has resolved. He told me he had been treated for syphilis in the late 1940s and remembered weekly intravenous injections and infrequent buttock injections. His current syphilis blood test suggested the treatment had been successful.

In 2000 I submitted a letter to a journal documenting the treatment for syphilis at St Marys under the care of the three consultants working in my department in 1948. Even though penicillin had been introduced a few years previously, the older treatments had not been dropped, so that patients received arsenic intravenously for many weeks, intramuscular bismuth which left tell-tale signs in the buttock area as documented above, as well as penicillin every three hours.

Bismuth was not as effective as arsenic. It was injected in the form of a cream into the buttocks at infrequent intervals.

Arsenic has been described as 'the poison of kings', and was used by the emperor Nero to poison his stepbrother Britannicus in order to become Roman emperor. It is widely found in the earth's crust and in many everyday foods. Arsenic was developed as a medication by Paul Ehrlich and his numerous colleagues in Germany. They worked with hundreds of different arsenic compounds until they found one with maximal effect on pathogens but minimal toxicity. It was the 606th compound tested and was known by that name as well as Salvarsan. This 'silver bullet' enabled syphilis to be cured. In 1910 a total of 65,000 free samples were given away to doctors in all parts of the world. It became known as the 'the saviour of syphilis', but nevertheless could be toxic if not used carefully.

Penicillin dissolved in beeswax was also used as a repository injection in those early post World War 2 years. One can imagine, with a certain degree of horror, the process of warming the solid beeswax compound so that it became a viscous liquid, and the force needed to inject that via large, wide bore needles into the buttocks, to say nothing of the pain so produced. In the 1948 paper, written by the three consultants, using combinations of arsenic, bismuth and penicillin, treatment failed in 10 out of the 275 patients with early syphilis. Of course the 'failures' may have in fact been reinfections suggesting that overall the outcome was excellent.

Frith, J. Arsenic the 'poison of kings' and 'saviour of syphilis'. Journal of Military and Veterans' Health 2013;21:11–17.

Goldmeier, D. Tertiary syphilis. Sexually Transmitted Infections. 2000 Jun;76(3):222–3.

Men who were being treated for infected urethral discharges, i.e. urethritis, most commonly caused by gonorrhoea or chlamydia trachomatis (CT), would be told to refrain from intercourse for a certain period of time. In the case of gonorrhoea this makes logical sense because treatment failure is always a possibility. The NHS recommends intercourse abstinence until all partners are treated for CT, to prevent reinfection. But what about masturbation? The advice usually given to patients who have recently had CT or gonorrhoea is to refrain, the logic being that the inflammation of the urethra needs to subside, and that masturbation will prevent this. I am not sure there is any evidence for this notion. When I first began practicing venereology, as it was then, there continued to be moralistic overtones in the treatment of STIs. Insisting there was no masturbation post treatment was, I believe, a carryover from previous times when consciously or unconsciously there was a punitive addition to the antibiotics. That said, there are some genital conditions where sex or masturbation is encouraged. One of these is inflammation of the prostate gland, where no bacteria are found to account for the inflammation. This is termed chronic non-bacterial prostatitis. There is some evidence that by emptying the prostate regularly, the condition improves. In the past this was undertaken by a doctor digitally massaging the prostate via the rectum (the prostate can be felt on the front wall of the rectum). Whereas most men found this treatment painful, some enjoyed it and were turned on sexually. There is an anecdotal account of a man who continued to massage his prostate himself by standing on a chair and inserting a broom handle into his rectum. Unfortunately on his last foray into pleasure he slipped. The boom handle pierced his bowel wall and continued its journey northwards ending up in his lungs. He was found dead at the scene.

Chapter 11

THE POLITICS OF MASTURBATION

In January 1993 President Bill Clinton appointed Joycelyn Elders to be the first African American Surgeon General. In 1994 she was invited to speak at a United Nations conference on AIDS and was asked there whether it would be appropriate to promote masturbation as a means of preventing young people from engaging in riskier forms of sexual activity. She replied, 'As per your specific question, in regard to masturbation, I think that is something that is a part of human sexuality and it's a part of something that perhaps should be taught. But we've not even taught our children the very basics. And I feel that we have tried ignorance for a very long time and it's time we try education'. Following this statement, the conservatives in Clinton's government were outraged and she was asked to resign in December of that year. This should be seen in the context of Clinton's subsequent affair with Monica Lewinsky between 1995 and 1997, which involved various sexual acts including fellatio.

In an almost immediate response to these events The Lancet *published an editorial headlining 'The politics of masturbation'. It pointed out that mutual masturbation was the commonest form of sexual expression after vaginal intercourse in heterosexuals and went on to describe 'Judaic writers regarding it as a crime warranting the death penalty'. St Augustine and St Thomas Aquinas held similar views, seeing masturbation as both part of original sin and a 'sin against nature'. In the 19th century a physician in Michigan USA named John Harvey Kellogg felt any type of sex was detrimental to physical and emotional health – hence he kept separate bedrooms from his wife and never consummated their marriage. He felt 'self pollution' was a 'crime doubly abominable'. He went on to hold views that a good wholesome non meat diet was the antidote to developing sexual passions – hence his development of the eponymous cornflakes. Contemporaneously a number of devices came on to the market to prevent men from touching their genitals e.g. a genital cup with built in padlock or a penile ring with spikes.*

In a letter to The Lancet *in reply to this editorial I discussed the orthodox Jewish viewpoint on masturbation. The biblical prohibition is based on Onan's action, whereby he refused to make love to his dead brother's wife Tamar, so that he 'spilled his seed on the ground'. This can of course mean masturbation, but could very well have meant coitus interruptus. However semen loss in urine at both micturition and defaecation is a physiological event, likely due to overflow of fluid from the prostate. Such physiological events had previously been observed 2,000 years ago by the rabbis of the Mishnaic period. Orthodox Jews do in fact condone masturbation under certain circumstances, largely if it is within the context of what husband and wife do together for sexual pleasure or to assist in conception. Orthodox heterosexual Jewish couples do not have sex with each other for the length of the woman's period plus seven days after bleeding has ceased. Such sexual abstinence is followed by renewing the couple's physical intimacy with each other with the enthusiasm of a newly married couple. All that said, orthodox rabbis absolutely prohibit solo male masturbation that is not within the marital context and is not undertaken for medical reasons. Various orthodox Jewish sources recommend giving charity and fasting if masturbation has occurred. The death penalty is obviously not appropriate in the current era. Female masturbation is not so vigorously condemned, although it is not altogether condoned. Reform rabbis generally put no prohibition on masturbation. Orthodox Islam also prohibits masturbation. Some authors have based this banning on doubtful scientific evidence. In all three major monotheistic religions the prohibition is based on the biblical injunction as described above, and subsequent clerical directives rather than scientific evidence.*

Goldmeier, D. The politics of masturbation. The Lancet. 1995 Feb 18;345(8947):454. PMID: 7853971.

The politics of masturbation. The Lancet. 1994 Dec 24–31;344(8939-8940):1714–5. PMID: 7996995.

Patton, MS. Masturbation from Judaism to Victorianism. Journal of Religion and Health. 1985 Jun;24(2):133–46.

Rössner, S. John Harvey Kellogg (1852–1943): 'Masturbation results in general debility, unnatural pale eyes and forehead acne'. Obesity Reviews. 2006 May;7(2):227–8.

https://en.wikipedia.org/wiki/Joycelyn_Elders - cite_note-15

Yavaşçaoğlu, I, Oktay, B, Simşek, U and Ozyurt, M. Role of ejaculation in the treatment of chronic non-bacterial prostatitis. International Journal of Urology. 1999 Mar;6(3):130–4.

Speed, D and Cragun, RT. Response to 'Masturbation: Scientific Evidence and Islam's View'. Journal of Religion and Health. 2021 Jun;60(3):1668–71.

Chapter 12

BRAIN TRAUMA

In 1993 my second son Nicky was in his last year of teacher training at Hertfordshire University. He was on the last day of completing teaching practice at a junior school in Maple Cross. It rained heavily for most of the day. He was driving our old black Volvo 340. Poor visibility meant he did not see a large articulated lorry bearing down on him as he edged out of a T junction turning right onto a main road. The first that Tricia knew about the accident was a serious faced young police officer asking to speak to her at our front door. He told her that Nicky was unconscious, and had been taken by ambulance to Mount Vernon Hospital. Ever since then she has dreaded the sight of policemen anywhere near our front door. Unfortunately one of our daughters was involved in an accident as a passenger about ten years later. She fractured her pelvis, so that all her children's births had to be by caesarean section. Back to Nicky's crash. The car was a write-off, but because it was built heavily (the 340s were sluggish in acceleration because of this) he had survived what would otherwise have been a fatal crash. Tricia phoned me at work. I immediately dropped everything and drove to Mount Vernon. A friend drove Tricia there from home. The emergency team at Mount Vernon assessed his Glasgow coma scale as high, although a CT scan only showed heavy brain contusion and swelling but no internal bleeding. He was breathing spontaneously and had good blood oxygen levels but was nevertheless transferred by helicopter to the neurosurgical unit at Charing Cross Hospital, where an MRI scan showed a similar picture. By the time we arrived at Charing Cross the consultant neurosurgeon had seen Nicky, and prognosticated that he would make a good recovery. One of the family was with him day and night. Just under a week later he slowly began to regain consciousness, although he was confused as to his surroundings and confabulated that he was in hospital having had a bad asthma attack. The first words he spoke clearly were, 'Can I have a Weetabix?'

In spite of the severity of the accident, he was left with no long-term side effects apart from laughter that was prolonged for 2–3 minutes when he found something humorous – a not unpleasant sequela. Within a few weeks he was back teaching and went on to marry and have four children, although that marriage ended in divorce because of his wife's infidelity. The police decided not to prosecute him in view of the terrible visibility that day.

I unfortunately have seen the lasting effects of brain trauma on sexual functioning in patients under my care.

Brian was a 30-year-old man who had severe epilepsy, suffering 30 grand mal fits a day. Medication using many different combinations did not help. He had been seen at one of the best neurological hospitals in the UK, Queen Square London. It was therefore decided to perform psychosurgery. Prior to the surgery he had also suffered from bad generalised anxiety with an accent on fear of death. In terms of his sex life he was happy, having a high sex drive and frequent changes of female sexual partners as well as a regular partner. He was sent to our unit six months after the surgery because he now had no sexual interest. The referral had been instigated by his girlfriend. He was not really concerned that he had no interest in sex. We did a thorough hormone investigation on him, which turned out to be normal.

We later prescribed tadalfil (medication similar to sildenafil). If his girlfriend massaged his penis he obtained a good erection but there was no concomitant feeling of pleasure. We also tried mindfulness and sensate focus, encouraging him to focus on what sensations he felt when his penis was stimulated in the hope that sexual desire would follow. It did not. His girlfriend ended the relationship. Interestingly he now had a new job as an undertaker specialising in embalming, feeling no fear of death at all. His psychosurgery had involved parts of the cerebral cortex and the amygdala – both of which are involved in sexual desire and arousal.

Fred's (aged 35) history was rather similar, in that he had no sexual desire at all. However his trauma was a severe road traffic accident that involved brain injury. He was very happy to live with his long-term girlfriend, but post trauma had no sexual interest in her at all. He was an attractive man for whom sex pre injury was a very important part of his life. Again, he had no problems in obtaining an erection but found it gave him no pleasure. Hormone levels were normal. We tried some medications that might increase sexual desire such as low dose trazodone and bupropion, but there was no response. Fred only came to our clinic three times as he saw no point in reattending. His girlfriend however came a number of times, when she weepily told us how wonderful their sex life had been prior to the accident. She emailed me six months after her last attendance telling me she had split up with Fred.

Brain surgery for epilepsy is not commonly undertaken nowadays and there is very little in the literature on sexual side effects of such treatment. There was a vogue for using psychosurgery to treat sex offenders, although the outcomes were varied and the ethics of doing so have been questioned.

Traumatic injuries to the spinal cord and brain in men generally tend to produce erectile dysfunction and ejaculation problems in up to 80 per cent of severe cases. In 3 per cent there is an increase in sexual appetite.

It is possible that amygdala and its connections were involved in our two patients. That structure processes both fear and evaluates the emotional content of a sexual situation.

West, S, Nevitt, SJ, Cotton, J, Gandhi, S, et al. Surgery for epilepsy. Cochrane Database of Systematic Reviews. 2019 Jun 25;6(6):CD010541.

Baguley, IJ, Barden, HL and Nott, MT. Altered sexual function after central neurological system trauma is reflective of region of injury; brain vs spinal cord. Brain Injury. 2020 Dec 5;34(13–14).

Rieber, I and Sigusch, V. Psychosurgery on sex offenders and sexual 'deviants' in West Germany. Archives of Sexual Behavior. 1979 Nov;8(6):523–7. doi: 10.1007/BF01541419. PMID: 391177.1732–1740.

Altman, J. The woman who knows no fear. New Scientist. 17 December 1994.

Chapter 13

DRUG DEPENDENCY AND DEPRESSION; 1980s to 1990s

The 1980s and early 1990s saw the swansongs of some of the older style, 'I know best' consultants at St Mary's, who believed that the hospital world should revolve around them. Strategic decisions and forward planning had now devolved to the hospital management staff. The physicians (rather than surgeons, gynaecologists or paediatricians who had their own Divisions) met once a month at the Division of Medicine meetings where management and consultants debated the current and future planning of the various departments. I found the content of these meetings tedious, but was happy to just observe the various factions and cabals jousting with each other. At a more local level I arranged the weekly academic meetings in our unit. I was also happy to teach informally, that is with the patient in front of me. We had a constant throughput of medical students and postgraduates.

By the end of the 1980s my family had grown so that I now had a total of eight children. Tricia had curtailed her child psychoanalytical training at the Tavistock Clinic before completing the course. Her personal analytical sessions, which had been a requirement of her training, had been nothing short of disastrous. She lay on a couch for 50 minutes a day, five days a week, with the analyst stationed behind her. The idea behind such in-depth therapy is that the patient develops greater insight into the psychological mechanisms as they relate to themselves, their personality and fantasies. The patient is encouraged to report dreams and to freely associate whatever comes into their minds during the session. The analyst's role is to relate these thoughts and emotions to herself as the analytical mother, explaining to the patient by interpreting their unconscious meaning. This is termed 'working in the transference'. Tricia felt that she had been reduced to a 'needy, helpless child' in the sessions. It was difficult for her then to suddenly revert to a real life mother of many children. The whole process for Tricia was destructive. At her last session Tricia told the analyst both how she felt the analysis had been harmful, and that it did not sit well spiritually with her orthodox Judaism. That said, both psychologists I have worked with over the years had fruitful analyses, one with a Jungian analyst and the other with Dr Donald Melzer, a well-known Kleinian analyst and doctor. Melanie Klein herself had said, 'The very fact that the analysis gets into contact with deep-lying unconscious anxieties gives the patient the feeling of being understood and therefore revives hope'. Well, Tricia's analysis did nothing of the sort! We are both still angry with her analyst some 40 years later.

My father lived on for four years after my mother died. He suffered from ongoing depression with delusions that he was impecunious, inhabiting the cold, unclean and now emotionally empty former family home at Pennine Drive. He either stayed with my brother Michael or myself over weekends. He was a shadow of the person he had been before my mother died. On a Saturday morning in early 1983, while he was staying with me, he complained of chest pain. He was admitted to hospital but died two days later of a heart attack. He had financed a new factory to be built in London's East End. This and the house he lived in were sold, so that I was able to afford to move into a larger house in Edgware.

My parents are buried side by side in Bushey cemetery. The Hebrew inscriptions on their tombstones simply state their names and dates of birth and death. The burial grounds are close to the A41, a noisy main road. Single engine private aircraft buzz stridently just overhead, their student pilots flying upwind and downwind in great circuits around the cemetery. But then again, my parents' lives were never placid. I was too selfishly involved in my own problems to ask them about their lives in greater detail. As I gaze on the neat stones, I know their relationship was no love match. They were tossed like so much flotsam and jetsam on the unimaginable rough seas of Nazi Germany. Just two kindly yet flawed humans who felt a marriage of convenience and children were a better bet than sailing on alone. In my old age I think of them almost daily. May they lie in peace.

During the 1980s and early 1990s we took pleasant summer holidays in Dorset, Cornwall, Devon and North Wales, spending part of the vacation in rented cottages and some time in hotels. This usually meant driving in two cars. I particularly enjoyed the outbound journey, driving with the sun behind me into Hardy's and Betjeman's Wessex. No satnav in those days. Getting lost in spite of reading and rereading the paper map was part of the holiday fun. Wide vistas of greenery and blue sky washed over me in waves of excitement and unexplained but very welcome optimism. My children look back on these holidays with affection and halcyon glows of remembered childhood exploration and fun. That said, they were certainly 'townies', refusing to drink milk from cows on farms we stayed on, insisting that we buy 'real' milk from a carton instead. It did not seem to matter to them that it often rained or that beaches could be cold and blustery. These holidays were really more for the children than Tricia and myself. On a personal note I searched out local narrow gauge or preservation railways wherever we holidayed. The family rather reluctantly came with me on these steam-powered journeys. I recorded many of these vacations initially in silent 8 mm film and later on in video with sound. One of my sons, who was about eight at the time gave his opinion, captured for posterity, that the steam train ride was 'rubbish'.

In 1986 I attended a conference in Seattle. Inflight movies at that time were projected via ceiling mounted cathode ray tube projectors onto large screens placed on bulkheads. The inflight film on that occasion was *Back to the Future*. I was enchanted by that film, thinking about it long afterwards. Audiences generally consider it one of the best films ever made – so I am not alone in my opinion.

The taxi ride from the airport at Seattle to where the conference took place went past the vast Boeing factory – literally miles and miles of aircraft buildings.

In the late 1980s, going into the 1990s I became increasingly anxious and dipped into depression for days at a time. I still took codeine at weekends and sometimes during the weekdays in the evening. Abusing these opiates contributed considerably to my poor psychological state. I was really apprehensive, worrying about extra heartbeats, fearing that they implied something sinister. I saw my cardiologist colleague Rodney Foale who did an ECG and reassured me they were benign and related to my anxiety.

Matters came to a head in 1991. Days were peppered by multiple panic attacks, detonating unannounced like so many erupting volcanoes. Between these explosions of extreme fear, I fell into valleys of exhaustion and utter hopelessness. I approached Maureen Sheehan my ex-psychologist colleague. She arranged for my psychiatric admission under care of Professor Anthony Wakeling at the Royal Free Hospital in Hampstead. I lingered there in Nicol ward for eight months, being unresponsive to antidepressants. Each day in that ward brought untold misery and no hope of ever improving or anyone seemingly being able to help my desperation, isolation and lonely helplessness. The ward afforded no privacy. I was moved from bed to bed every few days, from one section of the ward to another, from paranoid mutterings of one fellow patient to wrist slashing in another, taking my handful of possessions with me a – toothbrush, half a bar of chocolate, a pair of spare underpants, a comb I never used. I ate almost nothing and looked skeletal. I became a terrified, melancholic and fearful vagabond. My hair and beard were long and overgrown. I became physically inert and felt worthless. It took all my courage and determination to consider even having a bath. I approached the nursing station to asking for a plug for the bath. There were none, so they gave me a polystyrene cup to lodge into the plughole. I was very often chided and criticised by the nursing staff who would repeat, 'you should know better as a doctor'. I feel to this day that the nurses, at this great teaching hospital, were uncaring – even punitive. Looking after a patient who showed no sign of improvement was no doubt demoralising for them. That said, they could have been kinder and less bullying. I seemingly distracted them from talking amongst themselves while sipping tea and coffee in the nurses' station. Ward rounds were conducted without me being present to put forward my case. Professor Wakeling hardly spoke to me. However, there was one compassionate healer there, Dr Geoff Halperin, a senior clinical psychologist. Even though I was too ill to undertake any specific therapy with him, his weekly sessions were a fount of compassion and understanding. Phoning family, who were a small island of hope, was very difficult as there were no mobile phones in those days and the one payphone on the ward was almost constantly in use. My older sons visited me in the evening. After six months it was decided I should have a course of electroconvulsive therapy (ECT).

I had met ECT before, but as a psychiatrist not a patient. At Friern hospital the placing of the electrodes on the temples, the amount of current in the shock and whether there was to be unilateral or bilateral treatment (that meant placing the electrodes on one temple or both respectively), was left to the junior doctor. The only expert on site was the anaesthetist, who induced unconsciousness and gave the intravenous muscle relaxant that prevented the musculature from responding with a classical visible fit, where the arms, leg and facial muscles would thrash about in uncontrolled fashion. The room where it was undertaken had the feel of an operating theatre, for which purpose it had doubtless been used in Friern's heyday. It had been noted in the 1920s that schizophrenics rarely suffered from epilepsy, but when they did have a full fit, the psychotic symptomatology markedly receded. It was left to the neurologist Ugo Cerletti in Italy in the 1930s to carefully research ECT, and to create dedicated machines for its use. There had been a fashion until the 1960s to induce fits by giving insulin, or epileptic inducing medications such as Metrazol. But the electric inducement of the fits won the day as the safest and most effective. It was initially given without general anaesthesia, as exemplified in One Flew Over the Cuckoo's Nest where it is utilised as a punishment. Nurse Ratched, who is the nemesis of the film's hero Randle McMurphy and a cold hearted tyrant, seems to be the nameless therapist who prescribes the ECT – without anaesthesia. No doubt that scene in the film represents the public's general conception of ECT. Nowadays a highly experienced, dedicated and compassionate team provide the treatment, which can be used to treat unresponsive depression, schizophrenia and bipolar illnesses. It is a remarkably safe and effective, sometimes lifesaving treatment, although patients do complain of transient short-term memory loss afterwards. Very rarely the memory loss may be permanent.

Clarke, P. Hip Hip Hooray, ECT turns 80! Australasian Psychiatry. 2019 Feb;27(1):53–55.

The thought of receiving ECT raised my anxiety to terror levels. The anaesthetic induction was by intravenous injection and not unpleasant. Emerging from the short, modified convulsion and anaesthetic was to enter a confused nightmare. Some semblance of being in touch with my environment and reality returned after about an hour. The treatment profoundly affected my short-term memory. In spite of all of this my mood improved considerably over about eight sessions, so much so that I ventured to a local chemist shop, wrote out a prescription for myself and restarted the codeine. Not a good idea, but rather the insane actions of an addict. My wife was understandably incensed and immediately told the ward staff. However, informing Prof Wakeling yielded no action. After a few weeks my mood relapsed slightly. I was therefore given two more ECTs to good effect and then discharged. I returned to work a few weeks after the ECT. However I was apparently too happy! My wife and Prof Wakeling decided I had a low level of mania, termed hypomania, in spite of my protestations that I was just not depressed any longer and merely content and productive at work (I was!). I was therefore put onto lithium for a few months.

My abuse of codeine accelerated. It necessitated me writing out prescriptions for myself and going to many different pharmacies around north west London so as to not raise suspicion in any one of them. I held a list of them in my work briefcase, as well as the dates that I had last attended. Sometimes, even 25 years later, feelings of guilt and fear arise spontaneously as I quite legitimately walk into pharmacies. During much of the 1990s I was driving under the influence of drugs and more than once nodded off on longer journeys, with my family in the car. Someone or something was looking after me – I could have killed them all but in the event did not.

Cossetted, compressed, euphoric delight
Other worldly hope, sunlit fantasy
Dressed as my lover, my sweetheart
Easy, sleepy, nodding … nightmare
Into the land of terror
No longer my darling companion
Enter the demon, slaying my soul.

In the summer of 1996 I managed to drive the family to a two-week holiday in Polperro in Cornwall. In spite of the idyllic surroundings I was morose. My world was dark and angry. The drug was now only keeping withdrawal symptoms at bay. I took the opiate every day all day. But I knew the game was up, and that at some stage I would have to stop.

I continued in this fashion till the end of 1996. I denied taking drugs to everyone, but like most addicts, my secrets were found out. I believe one of the pharmacies I visited phoned my GP. Likely I had been to that particular venue once too often, and the staff were concerned that all was not as it should be. I was told later that friends were suspicious because I appeared to doze off for a few moments at a time midway through conversation – a common phenomenon in active addicts.

When Tricia found out she was livid. She phoned the Royal Free. By this time Professor Wakeling had retired and his place taken by a new senior lecturer from the Maudsley – Scott Weich. When I saw him in outpatients he firmly told me that unless I came off drugs and stayed off them permanently he would inform the General Medical Council (GMC). Nowadays all doctors with alcohol or drug problems are reported to the GMC, fortunately in a less punitive fashion, and their progress monitored closely.

I told my colleague Willie Harris the whole story. He discussed it with the medical director David Mitchell. They were both very supportive.

Rather than being sent to a specialist unit for addiction, I was readmitted to Nicol ward at the Royal Free and put on a tapering dose of codeine, arriving at zero after about ten days. Being opiate free was physically and emotionally overwhelming. Exhaustion led to dozing off, which led to panic attacks. Severe muscular pains, shaking limbs, explosive diarrhoea and background anxiety that trumped whatever I had experienced in the past, filled 24 hours of every day. The ward again felt like a living hell, with all the staff and patients taking on the role of Chiron ferrying me to an opiate withdrawal Hades. I felt utterly devastated and abandoned. The bed-moving saga continued as per my previous admission. A general psychiatric ward was the wrong place to withdraw from opiates. Ideally I should have done this at an addiction centre. Tricia would not talk to me, and insisted I sign our joint bank account over to her named account. She would not answer my phone calls. I discharged myself after a few weeks still feeling horribly edgy, aching and hopeless. I was shunned and ostracised by Tricia and spent most of the day at home on my bed in isolation. Tricia felt she had been duped and lied to by me, which was of course spot on. She became a spouse in name only. She only communicated with me to sort out factual details of common concern to us. Nevertheless, as the mother of our children, she felt she should give every assistance in my recovery. In retrospect she showed great loyalty in spite of despising me. She contacted the British Medical Association's sick doctors group who advised that a four-week treatment programme was available, and that the cost for this, which was considerable, could possibly be met by a local council grant. I was assessed as to my suitability for such a residential programme by an addiction counsellor in Colindale Hospital who forwarded his report on me to the local authority. I was eventually told that 90 per cent of the fee would be paid for me. Nicky my second eldest son took me to Foxleigh Grove in Maidenhead, which was a care home for the elderly. However, a section of it was devoted to therapy for addictions and was led by a psychologist/counsellor by the name of Rory O'Connor assisted by three junior counsellors, all of whom were recovering alcoholics or addicts. The unit specialised in the treatment of doctors, dentists and pharmacists. A local GP assessed me physically and took some blood tests. It was certainly an improvement on the Nicol ward, with pleasant surroundings and décor. The therapeutic day began quite early at 8 am and went on to 5 or 6 pm seven days a week. I was still having multiple panic attacks, particularly in the morning. I was too weak to run. Walking for an hour each day before breakfast did not really help. For most of the day we worked in groups. Some very senior, hitherto highly respected doctors, attended the course with me. Many had serially relapsed. A pharmacist joined who had helped himself to whatever medication he desired in his pharmacy. He was prepared to stop all of that, but could not conceive of life without cannabis. I left Foxleigh before he did. I did not stay in touch. I hope he did well. A dentist arrived at the same time as I. He had become addicted to the nitrous oxide (NO) gas that is given to dental patients to reduce anxiety and pain, and is given with oxygen. However he used the NO neat and sometimes for hours at a time. At weekends he replaced it with whiskey. He made an excellent recovery.

The basis of Foxleigh's treatment was the 12 step programme, originally devised for Alcoholics Anonymous. Anger rose up inside me, as a companion to fear. Anger at how I had ever got myself into the position of being an opiate addict. Exasperation that life had dealt me such a ghastly deal. I cheekily asked, 'What is the evidence that the 12 step treatment works?' I was quietly but firmly told that I should not be questioning efficacy.

The 12 step programme's basic message is that an addict (or alcoholic or a indeed person with any addiction) can't control their dependency on their own. Only a higher power can give the strength to do so. Past errors are detailed and shared with a person who has been through the steps. This 'sponsor' takes the addict through the steps. This leads to a new life with new codes of behaviour and helping other addicts. The original AA programme was constructed by Bill Wilson, a failed lawyer, and Dr Robert Smith ('Dr Bob') a family doctor, in 1930s Akron Ohio. Together they formulated the 12 steps, taken from Edwin Thacher's earlier Oxford group, which proposed six tenets for recovery based on Christian spirituality. The 12 step programme was devised to be spiritual, not religious. Objective evidence that the 12 step programme works is hard to come by, in part because it is difficult to find a suitable control group. In 2020 John Kelly and others published a report of 27 randomised controlled studies (over 10,000 patients) comparing the abstinence outcome of the 12 step programme to cognitive behavioural therapy for alcoholics undertaking the 12 step programme. The 12 step programme had marginally better outcomes at 12, 24 and 36 months. It is certainly better than trying to treat addictions solo.

Kelly, JF, Humphreys, K and Ferri, M. Alcoholics Anonymous and other 12-step programs for alcohol use disorder. Cochrane Database of Systematic Reviews. 2020 Mar 11;3(3):CD012880.

I wrote about each of the first five steps at Foxleigh. Step 1 – I was certainly powerless over codeine and indeed my life had become unmanageable. Step 2 – a power greater than myself proved problematic, but in the end I chose this power to be the Narcotics Anonymous (NA) programme. Step 3 was turning my will over to this higher power. Step 4 was a searching and fearless moral inventory which I wove around my life story, and step 5 sharing this with someone plus myself. These steps were shared with one of the counsellors in one to one sessions. The group was of course ever changing, with addicts and alcoholics coming and going. Group discussion topics were what was common to all, such as 'relapse warning signs' which again I challenged in my anger. Every evening we went to either an Alcoholics Anonymous or Narcotics Anonymous meeting in the local area. One of us drove the Foxleigh minibus. As I did not have a problem with alcohol I could not, with my hand on my heart, say I was an alcoholic in AA meetings. This was not received well in the rather middle class AA meetings in and around stockbroker belt Berkshire. I was told in no uncertain terms I was in the wrong place. I have not been back to an AA meeting since leaving Foxleigh. I hope AA are more understanding and welcoming nowadays. Afterall, alcohol is ethyl alcohol, a drug, and as such a substance of abuse. The addiction process is common to all substances of abuse.

Once a week we each had a meeting with Rory O'Connor, who had separate meetings with Tricia. I guess he did not intend to make me feel guilty about the effects my addiction had on my family, but that was the result. By the end of the four weeks' treatment I felt I had achieved something. I don't think the staff at Foxleigh held out much hope for me having a long-term recovery. I returned for short sessions every week or two for three months, where it was deemed I had made good progress. By this time I was going to about five NA meetings a week. I am still unsure if doing the steps was useful for me. But certainly meetings were a focus, and I appreciated the fellowship of other addicts and learnt something from every addict who shared their story. I completed working the 12 steps of NA with a 'sponsor' who I shall call Bob who was a lawyer who had been in recovery for about five years. It took me about three years to complete the 12 steps.

Later steps involved me attempting to make amends for the havoc and destruction I had wrought on Tricia and my family. Tricia attended 'Families Anonymous' or FA, where the friends or relatives of addicts could go through a 12-step recovery programme of their own. Their first step is 'We are powerless over our addicts', meaning the spouse, friend or partner has no control over whether the addict takes drugs or not.

It took about a year of recovery for Tricia to become friends with me again, and about ten years for her to feel she could be a loving wife.

I left Foxleigh in mid May 1997. While there I stayed up late to see Tony Blair win the General Election by a landslide majority. My stay and attending NA meetings was transformative. I felt more part of my family, less anxious and keen to return to work. We holidayed in the summer of 1997 in west Wales at Haverfordwest. I went to NA meetings in the wider locality, sometimes journeying for 90 minutes to meet up with just one other addict. Being drug free had many benefits, including not having to roam around London in order to find opiate dispensing chemist shops.

My unit director by this time was Dr Libby Claydon, Dr Harris having taken on more responsibility in the wider hospital. I phoned her from a local phone box to hear that I could restart work on a part-time basis in the September.

The end of August brought the tragic news of the death of Princess Diana. I and a number of my children waited for her funeral procession to travel past a bridge over the M1, where we paid our respects.

Chapter 14

POST RECOVERY YEARS

So in September 1997 I re-embarked on my career. It was a watershed moment. Life without opiates meant having to confront my mental demons without a pharmaceutical crutch. It was also an emotional rebirth, after 24 years of blunted emotions and spiritual anhedonia. I continued my almost daily cardio exercise, running on streets, which would take a toll on my body in years to come. In the first year back at work I felt a pull towards meditation, wondering if an understanding of and practising the basics of meditation would help my recovery. I was enticed by the notion of spiritual enlightenment after which there would be no suffering, merely pain and naked anguish. I rooted around for some months and eventually found one John Piper, who taught English at a major English public school. He had spent 12 years at the feet of a master Guru in India. He was happy to initiate me. In fact I had been introduced to Transcendental Meditation some 20 years before, but had not taken to it or carried on with its practice. John was middle aged, short, stocky, with unruly black hair and was always short of money, often coming to my house hungry. I supplied him with food and advanced payment for future sessions. In spite of his shortcomings he was a compassionate and insightful man. He taught me that the mind is disorderly and that when I was trying to concentrate on the breath, it was pretty normal to become distracted with thoughts and images – the so-called 'monkey mind' flitting from ideas to images, all in a trice. Noticing these events and feelings, without getting involved in them, but rather as if saying, 'oh I see such and such a thought' and just letting it float in and then out of my conscious mind was quite revealing and at times could be fun. I am not sure it directly helped the anxiety but at least I felt I was on some sort of road to a greater understanding of myself. John disappeared and reappeared in my life rather like the undulations of a yo-yo. No explanations given. I was always happy to see him. He told me his wife was quite seriously sick but did not elaborate. At about this time Nicky my second eldest son, aged 28, and otherwise hitherto quite well, was diagnosed with a cancer of his nasal sinuses. Treatment involved radiotherapy, which burnt his mouth, chemotherapy and lastly radical facial surgery. He had a recurrence in his lymph glands a year later (but has been well with no further recurrence over the last 22 years). John always had time to speak to him, not asking for payment. He would go for long walks with Nicky (who he called 'the sick one') who found them inspiring and reassuring. Then as suddenly as he came into my life he vanished. Gone. I have never been able to trace him. All that comes up on a Google search is John Piper the theologian, or a man of the same name who is an artist. I often think about him and wonder if he is still alive.

My clinical psychologist colleague Agnes Kocsis was also very interested in meditation, but for her it was a potential new therapy, to add onto her expertise in cognitive behavioural therapy. In the USA Jon Kabat-Zinn, who had trained as a molecular biologist, had started a mindfulness based stress reduction programme in 1979, having previously trained in Zen Buddhism. To reiterate, mindfulness is insight meditation, where one notices what is happening in a non-judgemental way from moment to moment rather than concentration-based meditation which pins the mind to one subject such as a mantra or gazing at a candle flame. By the end of the millennium mindfulness had been taken up by clinical psychologists such as Mark Williams in Oxford, building on Jon Kabat-Zinn's work, resulting in a new wave of therapy – mindfulness based cognitive behavioural therapy. Agnes attended one of his training courses at Oxford. I undertook an online course and read and reread Mark Williams's self-help book using mindfulness based CBT *Finding Peace in a Frantic World*. I did at least 30 minutes daily of one type of mindfulness or another for 20 years. I came to see how useful it was in terms of sexual therapy. I am not sure that it was therapeutic or indeed helpful for me in any way, but more of that later on.

Going back to work, and being drug free, gave me a new lease of life. I did not know it at the time but I had 23 very productive years ahead of me.

Chapter 15

SOCIETY FOR SEX THERAPY AND RESEARCH (SSTAR)

In April 1998 I attended a meeting of SSTAR, an east coast US academic group that ran annual conferences. I had previously been to one in New York City in 1995 and enjoyed it. The 1998 meeting was in Miami. I was excited at the notion of spending a few days in the USA. A British Airways' Boeing 747 transported me there – a wonderful aircraft, now alas obsolete as a passenger plane. I gazed out of the plane window, high above the Atlantic in a window seat. Unfortunately I am no longer able to occupy these grand vista positions as my elderly enlarged prostate translates into annoying my fellow passengers by my frequent visits to the toilet. Sky gazing was, and still can be, a powerfully engaging experience. The memories of my departure world, its buildings, colleagues, family and friends, fell away as the ground receded. Concerns about landfall, border control and luggage retrieval shrank to the miniscule. We were cruising at 37,000 feet. On that flight I was fortunate that the seat next to me was vacant and I could stretch out. The four Rolls-Royce RB211 engines orchestrated a relaxing hum. The cabin airflow resonated in the background. All was well in my mind. I felt my limb tension ease. I look around inside the cabin. A lot of people. And they have brought along tons of luggage. At take-off the 747 will have been loaded with 200 tons of aviation fuel. So how does this magnificent machine stay aloft? Its two mighty wings. We are told that the airflow over the curve of the top of the wing produces a partial vacuum, and hence lift. The exact mechanism still appears to elude flight theorists. Even Albert Einstein had a go at explaining it in 1916. In his later years he criticised his explanation as 'youthful folly'. Cruising at high altitude, a vast carpet of white, off-white and light grey cloud lay below, and a ceiling of dark blue above, diluting into white at the horizon. Just there. Just then. Just me. Tabula rasa.

There is now good evidence that viewing blue light enhances mood.

Miami airport was busy, as it was a connecting hub from Central and South America. The sombre woman at Miami border control lazily asked the reason for my visit. She broke into a broad smile when I told her I was attending a sex therapy conference. Miami was warm and sunny while the UK had an unusually wet and windy April 1998. The five-hour time difference meant I had a very long pre-conference first day, during which I walked along a sun-drenched promenade. The conference venue was held in a very pleasant four-star hotel. The bedroom was sumptuously laid out, with a large television and king-sized bed elevated on blocks, so one literally had to 'climb' into bed. As I unpacked my suitcase I remarked to myself that I had come a long way, physically and metaphorically, from an inpatient hospital bed in a locked ward, to a luxurious venue where I was free to come and go as I pleased. Over the years I got to know a number of the leading players in the US sexual dysfunction world.

Professor Yitzhok (Irv) Binik holds a chair in psychology at McGill University in Montreal. Irv at that time was a tall, slender, man with thinning brown hair and a trimmed beard. He could have been mistaken for a rabbi, and indeed had spent some time as a young man at a Jewish Theological Seminary. Over the years he developed a slight stoop, seemingly the result of chronic lower back pain. When talking one to one, his ample smile lit up his face. His expertise was pain at vaginal penetration.

Over the ensuing 25 odd years, Irv promulgated the idea that two conditions that arise in women, namely vaginismus and provoked vestibulodynia (PVD) are not distinct entities, and that there is much overlap between them. In the various updated national and international classification of diseases, these conditions are now classified into one overarching unified term – female genito-pelvic pain disorder (FGPPD). Before I discuss these conditions in detail, an aside to explain FGPPD by using colours as an analogy. Lime green and olive green are separate colours. One can classify them both as subdivisions of green. Any one colour could, say, be more lime green than olive green. An artist would know the two ends of this green spectrum, and will make up the exact shade he requires. In terms of classification the purists can delight in saying both can be named green. But for the artist, the exact shade must be known and accurately reproduced. So it is with vaginismus and PVD. They can be classified under the term FGPPD. However, treatment must be based on what the clinician finds at examination. So let me try to define these two conditions as simply as I can. Firstly vaginismus.

So, what is vaginismus? At its simplest it is inability to allow penetration into the vagina because of tense, contracted pelvic floor muscles. A more detailed definition is 'The persistent or recurrent difficulty of a woman to allow vaginal entry by a penis, finger, or object, despite the woman's express wish to do so. These women also tend to avoid intercourse, experience involuntary pelvic floor muscle contraction and anticipate fear, or experience pain'. I have already discussed Christine's case. Treatment may be as straightforward as explaining how to use vaginal trainers, along with a greater or lesser degree of psychological support. Such treatment may be given successfully in group format. For those wishing to have anal sex, anismus may be treated in similar fashion. A number of factors may predispose women to vaginismus. These include excessive religiosity, with negative views on sexuality and sex prior to marriage, past sexual abuse, the trauma of gynaecological examinations at the hands of unsympathetic healthcare workers and a background of generalised anxiety or phobias. Physical factors may also contribute to the onset of vaginismus. These include lack of sexual arousal (dry vagina), vaginal infections such as thrush, medications such as older types of antidepressants, post-menopausal vulval and vaginal changes which cause contraction of these tissues and lack of lubrication and vulval skin problems.

Eshana was a 33-year-old accountant who was brought up in the USA, emigrating to the UK when she was 20. She had recently dated a man she liked, and enjoyed kissing him. However when he attempted to touch her vulva, she became highly anxious and agitated, and withdrew from the embrace. She had never self-stimulated or explored herself when younger and had used sanitary towels rather than tampons. Eshana felt reasonably comfortable being examined by a female colleague. The area between the vulva and vagina appeared normal.

She was found to have tense and tender pelvic floor muscle as a well-lubricated examination finger was inserted into her lower vagina. The examining doctor also reported that Eshana looked fearful as the examination took place. The next session involved explaining simple genital anatomy and physiology. At the subsequent session she allowed a female physician to put a finger inside her vagina to a depth of about 3 cm. She was then asked to tense and then relax these muscles, and found, to her surprise, that once relaxed, the examining finger no longer induced pain. She was then given a set of plastic vaginal trainers, and told to contract and then relax her pelvic floor (this results in overall relaxation of the muscles) and then insert the smallest trainer and leave it for 30 minutes. Once able to insert and keep this trainer inside, she progressed to the next size up. After four weeks she was able to insert the largest trainer without discomfort and had started to self-stimulate clitorally while it was inserted. When seen a year later she was in a new relationship and was enjoying penetrative sex.

Priscilla was a 50-year-old post-menopausal solicitor who had not been in a sexual relationship for nine years. She had more recently started a new relationship with Claudine and had surprised herself by being very sexually responsive. However vaginal penetration with sex toys was not possible – dildos would not go in and trying was painful. After an examination I explained to her that the menopause had caused contraction of the vulva and vagina. I reassured her that these changes could be safely reversed using vaginal oestrogen cream. There was some tenderness of the pelvic floor muscles on using one finger to examine her, which I explained would not be troublesome after the post-menopausal changes had been reversed. She went away to discuss her treatment with Claudine. In the event they decided that penetrative sex was not necessary for them to have enjoyable sex.

The other condition is provoked vestibulodynia (PVD). Dynia refers to pain. The word is derived from the Greek meaning psychological pain or sorrow. 'Vestibulo' means in the vestibule. But where is the vestibule? It is the small, ring-shaped area between the inner labia (inner lips) and the vagina. At its simplest, PVD consists of tender areas in the vestibule, thus making penetrative sex painful. It had been more accurately defined by Friedrich in 1987 as having three criteria which I still find useful.

1. There should be severe pain on vestibular touch or attempted vaginal penetration.
2. There should be a physical finding of redness in some areas of the vestibule.
3. There should be tenderness or pain localised to certain areas in the vestibule when touched with a cotton bud swab.

The tender patches in the vestibule are usually red, but may appear normal. Their microstructure is varied but may show inflammation and immune activation, new blood vessel formation and in some cases excessive growth in the nerve endings. It may be associated with having had a candida (thrush) infection in the past, having been on combined oral contraception (meaning oestrogen plus a progesterone) and stress. There may be reflex contraction of the pelvic floor muscles, thus making the distinction between PVD and vaginismus difficult.

Treatment of PVD includes local anaesthetic cream used on the tender areas prior to sex, antidepressants used for their anti-neuropathic (anti-pain) quality, such as amitriptyline or duloxetine, certain antiepileptic medications which alter pain appreciation, such as pregabalin or gabapentin, pelvic floor physiotherapy, mindfulness based cognitive behavioural therapy, and in some cases injections of botulinum toxin (botox) or vestibulectomy (excision of a horseshoe part of the vestibule which ranks as day surgery).

Marlene was a 28-year-old secretary who gradually developed pain at vaginal penetrative sex over a number of months. She had been on the combined oral contraceptive pill for six years. She enjoyed foreplay with her partner but found penetration caused quite severe knife-like pain in the lower vagina, which lingered as a dull ache for hours afterwards. Examination showed she had red areas in the vulvar vestibule which caused moderately severe pain (7 out of 10 in intensity) in a number of sites on touching the red areas with a cotton bud. There was some mild contraction and tenderness of the pelvic floor muscles. She was diagnosed as having PVD. She changed her contraception to a copper IUD and was given lidocaine anaesthetic cream to use on the painful areas 30 minutes before sex. After spending more time becoming sexually aroused she was able to enjoy penetrative sex with minimal discomfort. A year later she no longer needed the cream.

Crowley, T, Richardson, D and Goldmeier D. BASHH Special Interest Group for Sexual Dysfunction. Recommendations for the management of vaginismus: BASHH Special Interest Group for Sexual Dysfunction. International Journal of STD & AIDS. 2006 Jan;17(1):14–8.

Crowley, T, Goldmeier, D and Hiller, J. Diagnosing and managing vaginismus. British Medical Journal. 2009 Jun 18;338:b2284.

Parish, SJ, Cottler-Casanova, S, Clayton, AH et al. The Evolution of the Female Sexual Disorder/Dysfunction Definitions, Nomenclature, and Classifications: A Review of DSM, ICSM, ISSWSH, and ICD. Sexual Medicine Reviews. 2021 Jan;9(1):36–56.

Paavonen, J and Eschenbach, DA. Localized Provoked Vulvodynia-An Ignored Vulvar Pain Syndrome. Frontiers in Cellular and Infection Microbiology. 2021 Jun 17;11:678961.

Friedrich Jr., EG. The vulvar vestibulitis. Journal of Reproductive Medicine. 1987;32:110–14.

Ekström, JG and Beaven, CM. Effects of blue light and caffeine on mood. Psychopharmacology (Berline). 2014 Sep;231(18):3677–83.

Peggy Kleinplatz, who is a professor of medicine and psychology in Ottawa, gave inspiring pre-conference talks on 'Optimal Sex' at SSTAR meetings. She is small in stature, but monumental in knowledge and exposition. At a meeting in the new millennium she gathered a quorum of Jewish men at a meeting so that she could recite 'Kaddish', the Jewish memorial prayer for her father who has passed away some months before. Both her parents were Holocaust survivors. Her mother Sarah passed through many concentration camps and was finally liberated by the British in the Bergen-Belsen camp. In a video Sarah attests to the fact that at her liberation there 'was more crying and mourning than joy'. Reverend Leslie Hardman, who was the British Army Jewish padre for the liberators, found the experience of liberating the survivors equally daunting.

A woman, seeing the Jewish Star of David on his uniform, came forward to embrace him. His instinctive reaction was to recoil in shock and disgust from this skeleton of a woman covered in skin. He steeled himself almost immediately, and reciprocated the embrace. Reverend Hardman came to my school on the annual Remembrance Day to lead Jewish prayers for the school alumni who had fallen in the World War. Little did I realise then that I was listening to a man who had seen and experienced so much horror.

Freda and Dwayne were a couple who had been together since their 20s and both were now nearing their early 70s. I originally saw Dwayne because his penile glans had become very red and sore. He was found to have candida (thrush) which was likely underpinned by type 2 diabetes. On further questioning he told me that he had been unable to obtain an erection for the previous two years. I eagerly told him that I could remedy that with medication. He gently told me that he did not really want a quick fix for a mechanical problem and that tablets were not necessary as Freda and he were having very pleasurable sex without the need for an erect penis. I became interested in what was happening in their sex lives, so I asked to see both of them to learn more.

Seemingly in the early days of their relationship a hard penis and frequent vaginal sex were de rigueur for them, fitting in orgasms between their busy work schedules and bringing up their children. Things gradually changed so by their 60s they had more leisure time from work and their children had now left home. They both went on a mindfulness course followed by a number of month-long retreats. Physical intimacy gradually developed into a more spiritual event and they talked more about how their bodies and minds felt when touching each other. They learnt more about their partner's inner sexual and emotional world and were happy to share this with each other. Their intimate sessions could go on for up to two hours and sometimes this culminated in an erect penis but often not. They told me that their newfound physical satisfaction was far greater than the sex they had in their youth. So when Dwayne developed erectile problems it was not a real issue for them.

Peggy Kleinplatz conducted phenomenological orientated research from such 'experts', who see themselves as great lovers, to determine the components of optimal sex. She and her colleagues recruited individuals who responded to their request for persons who experienced 'great sex' to contact them. The average age of the older individuals in their study was 66 (range 60–82). Ten of them had significant illnesses or disabilities. The lessons learnt from these 'experts' were:

1. Great lovers are made, not born. They often had to unlearn the sex they had while growing up and had to develop a new personalised vision of sexuality. This meant that although new sexual journeys might be risky, they were perceived as very real and raw and they, the 'experts', were better able to communicate and listen to their partner's sexual needs.
2. Optimal sexuality flourishes in the context of a relationship deepening with time.
3. They were less willing to settle for anything other than what they really wanted and therefore sought appropriate partners.
4. Great sex takes a lot of time. Some couples spent many hours in play that gradually involved sex.
5. Exploration and familiarity both have advantages. Novelty is not just about new positions and sex toys but rather exploring sex at a deeper level.
6. Age, illness and disability are not necessarily obstacles to optimal sex. The perception of many of the 'experts' was that they were more able-bodied than their medical diagnoses reflected. One quote was 'you can always find sex in a new form'.
7. Ability to say 'no', but rather only to engage with the partner if both wish to explore sexually and to discover mutual paths of discovery.

Bernie Zilbergeld, who I never met but did come to much earlier SSTAR meetings, was an unusually insightful psychologist who practised in the West Coast of the USA in the latter part of the 20th century. He debunked some cherished male 'sexual myths'. Such false ideas included notions that 'a man is always ready to have sex, penile erection is necessary for female sexual pleasure and that the penis has to be rock hard all throughout sex'. He added to our understanding of what normative male sex should look like and gained prominence, some would say notoriety, by challenging the work of William Masters and Virginia Johnson. Writing in *Psychology Today*, Zilbergeld described Masters and Johnson's research as 'flawed by methodological errors and slipshod reporting'. There has been a tendency for these male sexual myths to be perpetuated and enhanced with the advent of medications such as Viagra.

Kleinplatz, PJ. Optimal erotic intimacy, pp 318–330. In Handbook of Clinical Sexuality for Mental Health Professionals. 3rd Edition Ed. Levine SB, Risen CB, Althof SE. Routledge Taylor and Francis, New York, 2016.

Zilbergeld, B. The New Male Sexuality. Bantam Doubleday Dell, New York. 1999.

In the early 2000s, Professor Stephen Levine, who holds a chair in psychiatry at Case Western Reserve University in Ohio, gave a number of half day pre-conference seminars at SSTAR meetings on the topic of love and infidelity. I have already given reductionist explanations of romantic love above. Steve's sessions discussing love were tantalising forays into the mind of a scientific thinker who was at the same time a master wordsmith. There was poetry in his explanatory prose. 'Love is both a noun and a verb, a thing and a process, a concept and a set of actions. It is an ambition, an intrapsychic bond, a social arrangement, a personal commitment, a management process, a force in nature, an illusion, a set of emotions and a stop sign'. 'Love contains the eternal human quest for oneness, stability, harmony and happiness. Its processes include searching for love, falling in love, being in love, staying in love, losing love and mourning for its loss'. I suppose he was describing love rather than trying to understand its scientific underpinnings. Steve was a keen and accurate observer of the human condition. A latter day psychodynamic Shakespeare! At one seminar he was asked about rekindling intense romantic love by a person having an affair who was already in a gently harmonious long-term relationship. He thought for a good minute, then with a smile, both whimsical and sagacious, he told us that such a path would be pleasurable but ultimately destructive.

Levine, SB. Barriers to loving: A clinician's perspective. Routledge, New York, 2013.

I first met Sandra Leiblum in the early 2000s. Even before her seminal paper on persistent genital arousal she was a real star SSTAR member! Her energy, vibrancy and vivaciousness were legendary. By the time I knew her, Sandra was a professor of psychiatry at the Robert Wood Johnson Medical School in New Jersey. However, she had risen from humble roots in Brooklyn. She had an uncanny ability to foster intimacy and enthusiasm in those whom she met. We introduced ourselves over drinks at a SSTAR reception. She wanted to know all about me. I was fascinated by her twangy Brooklyn accent. Her lips seemed to do a lot of the enunciating. The Rs at the end of words became 'ah', so her 'fear' became 'fee-ah'. 'Thought' became a two syllable word as she pronounced it. As I was noticing her diction, she grinned widely and chuckled as a welcoming gesture. I congratulated her on her recent novel scientific publication. She replied with a 'yeah, but' and immediately launched into encouraging me and inviting me to collaborate with her into further work on persistent genital arousal. I was by now caught up in her whirlwind passion for research. Her lectures were spectacular, holding audiences spellbound in the palm of her hand. Yet she told me she was always highly anxious before any talk, and had to obsessionally go over the lectures beforehand. She died prematurely from a stroke and bike accident aged 67 in 2011.

So what is persistent genital arousal disorder (PGAD)? In essence it is a condition, usually described in women, where the patient complains of spontaneous arousal felt in the genital area, is unbidden and unrelated to sexual desire. A more exact definition has six features.

1. Subjective genital arousal, which may present as a feeling of fullness, swelling or sensitivity, that continues for hours and does not cease spontaneously.
2. These feelings do not cease with an ordinary orgasmic experience.
3. These symptoms are unrelated to any experience of sexual desire.
4. The feelings are usually spontaneous but may be triggered by sexual and non-sexual stimuli (e.g. a vibrating car).
5. They are unbidden, intrusive and unwanted.
6. They cause distress.

Sandra Leiblum, and her psychologist colleague, Sharon Nathan, first described persistent sexual arousal in a handful of patients in 2001. Up to this time women who complained of unwanted sexual feelings were probably thought to have excessive sexual desire, which was deemed to be underpinned by psychiatric states such as mania or illicit drug use. When presenting without an accompanying psychiatric illness, such women were likely diagnosed as having 'nymphomania'. Doctors in the 18th and 19th century (who were mostly men!) considered hypersexuality or 'nymphomania' to be part of the spectrum of 'female lunacy'. Physicians in the USA and UK not uncommonly advocated removal of the clitoris as a treatment for a number of such psychiatric states. Incidentally, one of the last English surgeons to advocate clitoridectomy for 'female hysteria' was Isaac Baker Brown who performed these horrific operations at St Mary's Hospital. He was later vilified by his colleagues.

What Leiblum and Nathan did was to actually *listen* to what the women were saying. Their patients did not have increased sexual desire at all – just unbidden arousal. Of course PGAD patients have been complaining of unbidden arousal for ever. The mind boggles as to how they were managed even in the latter part of the 20th century.

The first case of PGAD that I came across was in 2005. We wrote the case up in 2007. She had previously been admitted to another hospital because of very low blood pressure and was found to have two holes in the upper chambers of her heart (atrial septal defects) and was given a cardiac pacemaker because of a very slow heart rate. Since 2004 she had complained of spontaneous genital arousal for most of the day which was relieved temporarily by inducing multiple orgasms, produced by masturbation, sexual intercourse or a powerful vibrator. We tried a number of medications and psychological therapies but her PGAD remained unimproved.

Over the ensuing years I wrote a number of papers detailing PGAD with Sandra Leiblum and other colleagues. By the second decade of the new millennium, patients who likely had PGAD were searching the web to look for answers. Many of those in the UK and Europe alighted on my name after reading one or more of these papers and finally ending up seeing me, often via circuitous routes, having seen a number of perplexed and unhelpful doctors on their journey.

Some of Sandra Leiblum's original patients did not quite fit all the criteria for PGAD, even though they had spontaneous unbidden arousal. They were able to dismiss the sensations until an appropriate sexual scenario appeared e.g. waiting to get home to a partner. These women could be classed as having persistent genital arousal but not the full-blown disorder as they were not distressed by their symptoms.

The prevalence of PGAD in women has been estimated in clinical, university and national settings and ranges from between 0.5 per cent and 4 per cent. Our survey at St Mary's in 2008 of 100 successive women attending our sexual health clinic for a multitude of reasons yielded one patient with full PGAD.

There appear to be multiple possible associations with PGAD some of which might be causative. Studies suggest a significant association with anxiety, depression, obsessive compulsive illness (OCD) and sexual abuse. The mechanisms of production and perpetuation of the persistent arousal in the scenario of psychiatric illness might begin with mild short-lived spontaneous genital arousal which is normal in most women and akin to men's penises becoming somewhat engorged at various times of the day. Because of the underlying psychopathology e.g. fear and ruminations, the patient narrows their cognitions to the genital arousal which is maintained and augmented by the underlying anxiety.

PGAD resulting from starting, maintaining or withdrawing from SSRI or SNRI antidepressants have been reported a number of times in the literature.

Neuropathies, in particular small nerve fibres of the pudendal nerve and its branches (e.g. dorsal clitoral nerve also innervates the bladder) are likely causes in many cases.

Rarely problems at or near the spinal cord may be associated with PGAD e.g. Tarlov cysts near the spinal cord. A case has been reported of PGAD produced by eating excessive soya products, as has a case where there are large pelvic varicosities which resolved after the abnormal veins were treated

Treatment of specific causes is undertaken where possible. Mindfulness based CBT, antidepressants (not the SSRI or SNRI variety if this has caused the problem in the first place) or antiepileptic agents (both functioning as mood modifiers as well as benefiting neuropathic pain) and pelvic floor physiotherapy are the mainstay of managing these patients. That said, these patients are time consuming for the healthcare worker to manage, and their therapy can be exhausting and emotionally draining particularly when, as often happens, they make a poor response to treatment and continue to contact the healthcare workers on a frequent basis.

Leah was a 60-year-old single retired woman who had worked for the civil service in Wales for many years. She had been depressed two years before and had been put onto citalopram, an SSRI antidepressant. On this she improved considerably. Eighteen months before seeing me she had decided to slowly come off citalopram over a month. A week after she took the last of the medication she began to experience unbidden sexual arousal, mainly felt in the genital area. It seemed to come totally out of the blue, in that she had not been thinking about or viewing anything of a sexually arousing nature. The sensation lasted for many hours at a time. After stimulating herself to orgasm the sensations ceased for 30 seconds. They were improved by distraction and worse when sitting on a chair. At the same time she noticed some unusual urinary frequency. Her GP referred her to the local urogynaecology department where she underwent investigations, including a cystoscopy and MRI of her pelvis. These were all normal. She was discharged from their care with the recommendation to the GP that she be referred for a psychiatric opinion. The psychiatrists felt she was very anxious but could not explain her genital symptoms. Leah was becoming desperate by this stage. Her GP, although sympathetic, felt she could not help except for prescribing a tranquiliser (diazepam). Things took a turn for the worse at this stage as she began having multiple spontaneous orgasms when driving her car. Because of this she stopped driving. However the spontaneous orgasms were now happening many times a day. She felt ashamed, lonely and isolated. As no one seemed to be able to help, she began searching the web to see if her strange set of symptoms fitted in with anything. She found my name relating to PGAD and got a referral from her GP. When I saw her she looked terrified, but was able to excuse herself for not sitting down as it made her feel worse. I carefully examined her genital area and assessed her neurologically. All was normal. She was relieved when I confirmed her diagnosis of PGAD, which I felt in her case was precipitated by withdrawing from citalopram, possibly causing a problem with the pudendal nerve with innervates the genital and bladder areas. After four weeks of amitriptyline, a different type of antidepressant which also helps reduce pain by acting on peripheral nerves, most of her symptoms, including the urinary ones, had almost disappeared. An MRI scan of her spinal cord was normal.

Gael was a 17 year old working for her Higher exams in a rural part of Scotland. Every time she sat down to study she noticed genital arousal that seemed to come completely out of the blue. It throbbed and she also noticed spontaneous increase in vaginal discharge. Her family were practising Christians (Church of Scotland). When she was 11 she started worrying that when she died she would not be able to find her family in Heaven. A year later she went through a phase where she had to check the front door was closed. This ritual could take up to 15 minutes and sometimes made her late for school. When I saw her she had catastrophic ideas about the PGAD such as 'I will always have it every time I sit down to study'. All her investigations were normal. We started her on sertraline for her anxiety, which had a marked obsessive compulsive component. We also worked with her psychologically, teaching her mindfulness, as well as simple relaxation techniques. In her case, fear of the subjective genital arousal, as well as her tendency to obsessional thinking, had ramped up her normal background levels of arousal to PGAD levels. She was subsequently able to revise for exams with little trouble from the PGAD and did well in her Highers.

Claudine was a 68-year-old single woman who originally came from Northern Ireland. She had been troubled by spontaneous genital arousal for years but had not sought help till the feelings became continuous, along with a pain felt in the vulva. Her GP had investigated her thoroughly in terms of MRI scans and had started her on paroxetine, with her now taking a high dose. It did not help. At night in bed she suffered multiple spontaneous orgasm that often woke her up from sleep. She had contemplated suicide a number of times. As part of her workup when I saw her, she disclosed that she had been regularly beaten by her father as a child each time he came home from the Army. As her mother was chronically sick with severe depression she was taken into a care home where she was sexually abused multiple times, finally becoming infected with and later treated for syphilis. She recalled that the sexual assaults were accompanied by what in retrospect was genital arousal. She was intensely ashamed about this, feeling that the assault and syphilis were her fault – otherwise she would not have become sexually aroused. At a multidisciplinary meeting of our team it was felt that the current PGAD might well be exacerbated by her past memories. She had 25 sessions of EMDR (trauma therapy) addressing her past abuse. However this did not help her current symptoms. We added pregabalin, an antiepileptic medication that is useful for neuropathic pain as well as anxiety. Even at high dose it did not help. She was later referred to a pain clinic where a pudendal nerve block did not help. She still continues to have PGAD.

PGAD can of course affect men too. I managed a 24-year-old French PhD student whose life was ruined by having to masturbate to orgasm multiple times a day in order to obtain any semblance of a reasonable life. In spite of multiple medications, psychological treatments, nerve blocks and finally crowd funding a trip to the USA to have many of his pelvic nerves reimplanted, he is no better, living most of his life isolated at his parents' house.

Jackowich, RA and Pukall, CF. Prevalence of Persistent Genital Arousal Disorder in 2 North American Samples. The Journal of Sexual Medicine. 2020 Dec;17(12):2408–16.

Leiblum, SR and Nathan, SG. Persistent sexual arousal syndrome: a newly discovered pattern of female sexuality. Journal of Sex & Marital Therapy 2001 Jul–Sep;27(4):365–80.

Leiblum, SR and Goldmeier, D. Persistent genital arousal disorder in women: case reports of association with anti-depressant usage and withdrawal. Journal of Sex & Marital Therapy. 2008;34(2):150–9.

Goldstein, I, Komisaruk, BR, Pukall. CF et al. International Society for the Study of

Women's Sexual Health (ISSWSH) Review of Epidemiology and Pathophysiology, and a Consensus Nomenclature and Process of Care for the Management of Persistent Genital Arousal Disorder/Genito-Pelvic Dysesthesia (PGAD/GPD). The Journal of Sexual Medicine. 2021 Apr;18(4):665–97.

Facelle, TM, Sadeghi-Nejad, H and Goldmeier D. Persistent genital arousal disorder: characterization, etiology, and management. The Journal of Sexual Medicine. 2013 Feb;10(2):439–50. doi: 10.1111/j.1743-6109.2012.02990.x. Epub 2012 Nov 15. PMID: 23157369.

Michael Perelman is an Emeritus professor of psychology at Cornell Medical School. In 2010 he presided over the annual SSTAR meeting. The meeting that year was in Boston in April. It was crisp but not cold, and the city was showing off gorgeous magnolia flowering. He spoke in low pitched voice, almost basso profundo range. His expertise is delayed ejaculation, on which he is undoubtedly a world expert.

Benjamin was a 50-year-old engineer who had been with Sofie for four years. He had never been able to ejaculate at vaginal intercourse with her. He had similar experiences with previous partners. Solo self stimulation, particularly in the early mornings, would result in him orgasming and ejaculating after about five minutes. The problem was overcome to some extent by him masturbating with her being present and just prior to ejaculation inserting his penis inside her. They were not entirely happy with this, and at this point I lost touch with them. Two years later I received an email from them saying that Benjamin had penetrated her early one morning while she was asleep. In fact she was drowsy rather than being asleep, and welcomed him doing this. In other circumstances this could be seen as non-consensual sex. He was able to ejaculate after about three minutes. They continued to have early morning sex with him being able to ejaculate relatively easily. Levels of testosterone and cortisol peak in the early morning. It may be that Brian's level of innate sexual drive peaked at that time.

Phillip, an English student aged 19, came to our department because he was unable to ejaculate at vaginal sex with his girlfriend. He reported that that his penile sensation at vaginal sex was very different from sexual excitement and arousal he was used to. He felt that her lubricated vagina provided very little penile sensation. She was his first sexual partner. He had solo masturbated aged 13 onwards thrusting backwards and forwards against bedsheets whilst looking at porn on his phone. He was physically well and had no obvious psychological issues. We discussed the likelihood that he had become habituated to idiosyncratic self stimulation without lube whilst watching the porn. He agreed not to self stimulate or watch porn at all. After being abstinent for two weeks, he re-engaged with his partner and was able to ejaculate intravaginally. He told us that although he was initially still conditioned to dry solo sex, sexual abstinence enabled him to focus relatively easily on auditory, visual and olfactory sensory inputs when he was with his girlfriend. These overrode the low level of penile sensation. With continued sexual excitement restricted to sex with his partner, he was gradually able to appreciate and enjoy the feeling of penile thrusting in her lubricated vagina.

Friedrich was a 35-year-old married man who had never been able to ejaculate except in his sleep. He had been carefully assessed neurologically with no abnormality detected. His sex hormone profiles were also normal. He had been married for ten years and he and his wife had been unable to conceive. He had unsuccessfully used yohimbine (see below for details) and a powerful vibrator. He was referred to an andrologist who retrieved sperm surgically (aspiration of the epididymis, where the sperm are stored prior to ejaculation), injected the semen directly into his wife's cervix, with a resulting pregnancy.

Parker was a 45-year-old academic in a long term relationship with Samuel. He had no problems in terms of ejaculating until he was put onto an SSRI antidepressant (sertraline) after which he was unable to ejaculate by any means. His blood pressure was normal. He was given yohimbine to take prior to sex which resulted in him being able to ejaculate. On some occasions he had a drug 'holiday' – that is he didn't take the sertraline for a day or two which also helped him to ejaculate without the yohimbine.

DSM 5 (*The Diagnostic and Statistics Manual of the American Psychiatric Association*) defines DE as the inability to ejaculate during sexual activity at over 75 per cent of occasions in spite of adequate stimulation, specifically after 25 to 30 minutes of adequate stimulation. It may be lifelong or acquired and generalised (occurs in all sexual situations) or situational, for instance as illustrated above at vaginal sex but not solo masturbation. The patient needs to be distressed about the DE to make the diagnosis. There are men who can easily delay ejaculation for as long as they want, which may well be a source of pleasure for them and their partner. These men do not suffer from DE. DE is probably the least understood of all the male sexual dysfunctions.

Twin studies suggest that, although there appears to be a genetic component underpinning premature ejaculation, no such correlation has been found for DE.

True DE may be associated with a number of psychosocial issues. Religious proscription of premarital sexual arousal may result in conflicted attitudes towards sex after physical contact is permitted at marriage. Inadequate stimulation by the partner and fear of pregnancy may also lead to DE. A common cause nowadays is the early onset of masturbating solo, without lube, whilst watching porn online. This results in a level of friction and visual stimulation that may be disparate with the physical and psychic sensations of lubricated partnered intercourse. Some young men masturbate in an idiosyncratic way and become conditioned to sexual arousal by so doing e.g. by rubbing the penis against bedsheet also known as frottage. These scenarios need arousal retraining possibly at first via solo stimulation with lubricant but without online stimulation. The man will later graduate to do this with a partner if available.

There are number of medical causes of DE, including damage to pelvic nerves e.g. from diabetes or neurological conditions such as spinal cord injuries or multiple sclerosis. These patients may be able to ejaculate in their sleep.

Psychological treatments are detailed as above. Powerful vibrators placed on the frenular part of the penis can be helpful, particularly in order to retrieve semen for conception. Medications that may be helpful include yohimbine and ephedrine, both of which are adrenergic stimulants, that is they increase the level of adrenaline in the body, and imipramine a stimulatory tricyclic antidepressant. Drug 'holidays', for a day or two, particularly from antidepressants will not precipitate recurrence of psychiatric illness as it takes at least 14 days for them to be excreted from the central nervous system.

Jern, P, Santtila, P, Johansson, A et al. Evidence for a genetic etiology to ejaculatory dysfunction. International Journal of Impotence Research. 2009 Jan–Feb;21(1):62–7.

Martin-Tuite, P and Shindel, AW. Management Options for Premature Ejaculation and Delayed Ejaculation in Men. Sexual Medicine Reviews. 2020 Jul;8(3):473–85.

SSTAR meetings often had accompanying evening entertainment. One such event was the concert pianist Richard Kogan, also a professor of psychiatry, who gave a talk on Tchaikovsky's sexual and general psychopathology, illustrating these with various passages of his music played by himself. On other evenings one could socialise with colleagues or watch the then (unusual for the UK) multichannel TV. The usual night flight back (when I never could sleep) was followed by a few days of jet lag.

Chapter 16

THE EUROPEAN SOCIETY FOR SEXUAL MEDICINE (ESSM) AND PHOSPHODIESTERASE INHIBITORS (PDE5is)

ESSM meetings were held annually in various European capital cities in the late winter. In the early 2000s they were mainly attended by urologists interested in male sexual problems. We delegates generally had all our expenses paid for by a pharmaceutical company. The first ESSM meeting I attended was hosted by the urologist Hartmut Porst in Hamburg in 2002, where between sessions we strolled around the glittering Christmas markets, bravely chatting to our colleagues in the cold Baltic wind. I learned at the conference that Martha Bernays, Sigmund Freud's wife, was born and brought up in the city. My expenses that year were paid by Pfizer, who manufactured Viagra. I met Graham Jackson there, a London based cardiologist. He had taken an interest in erectile dysfunction (ED) some years before and pointed out the safety of intercourse in a man with a regular partner, where the amount of energy used up at man on top sex is the equivalent to a 20 minute walk or climbing two flights of stairs. Sex with a novel partner can of course expend much more energy and may not be safe for the cardiac patient. He also identified that ED of a physical rather than psychological nature is very likely to be associated with narrowed blood vessels in the whole body, as well as the penis. He recommended that all men with ED should be investigated to assess their cardiac risk. In later years Graham developed severe Parkinson's disease and alas is no longer with us.

By 2002 Viagra, or to give it its generic name sildenafil, had been prescribable in the UK for four years. The history of sildenafil, best known by its trade name Viagra, is worth reporting. Sildenafil treatment trials began in the early 1990s, the conditions under investigation being angina and high blood pressure. The side effect of enhanced erections in the men was a surprise to everyone, including the chief investigator Dr Ian Osterloh. In 1998 the Free Library described Osterloh as 'a shy soft spoken churchgoer, who drove a very ordinary car to his work at the large Pfizer pharmaceutical complex in Sandwich Kent'. As a result of this serendipity finding placebo controlled studies were undertaken treating men with ED. The results, showing sildenafil's great efficacy over placebo were initially published in 1996. I was involved in one of the early placebo controlled ED phase 3 treatment trials of sildenafil in 1996. It was obvious to me that half the patients did not improve whereas the other half made a marked improvement, even though I did not know what arm of the study any particular patient was in. My Belfast GU Medicine colleague Wallace Dinsmore was also at the Hamburg meeting. He and other investigators had found that giving nitrates, used for angina, together with sildenafil, could produce a catastrophic fall in blood pressure. He told harrowing tales over breakfast of patients in that study whose blood pressure was so low he thought they had died! The main financial sponsor of the Hamburg meeting was Pfizer. Not surprisingly, many of the research papers presented involved the use of sildenafil.

One of my first patients using sildenafil in a clinical setting, rather than any treatment trial situation, was a train driver working in the Western Region out of Paddington, which is just round the corner from St Mary's. He stopped using the medication because he found it hard to distinguish red from green, so making train driving a potential hazard. It is well know that the PDE5is affect the rods and cones in the retina. I advised him not to use sildenafil within 12 hours of train driving and indeed car driving. Following this advice his colour vision was no longer a problem.

In subsequent research, sildenafil was given to fit young men who had no problems with erections in order to obtain measures of the degree of hardness of the penis before and after the medication. A Rigiscan device was used, which consists of a small portable computer connected to two bands placed around the shaft of the penis. These record expansion and degree of hardness of the penis both lengthwise and circumferentially. Post sildenafil ingestion, the penis was harder than pre medication. Because of this, PDE5is have been used recreationally even in otherwise fit men without ED. The difference between erection hardness using and not using the PDE5is can give cause for concern and lead to psychogenic ED. On a more positive note, female partners of men with ED who used vardenafil, another PDE5i, reported increased sexual satisfaction after their partner's ED was resolved. This study describes an unusual medical situation, where a drug given to one patient results in improvement in health in another patient not taking the medication. Of course penile hardness is not the be all and end all relating to sexual satisfaction in both men and women. Nevertheless PDE5i can help men to obtain a second erection sooner after ejaculation than without medication. In many men this second (or third or fourth) erection soon after the first can last longer (to orgasm) than the first time round. This can be useful for some men with premature ejaculation, and for those women who value vaginal orgasm but tend to need a longer time of thrusting to achieve it. PDE5is have other beneficial effects in the body apart from penile blood vessels. These include treating acute mountain sickness and high blood pressure in the lungs – pulmonary hypertension. Their effects on the brain are illustrated by the case history elaborated below.

Bruce was a 65-year-old retired schoolteacher originally from County Cork. His wife gently nodded in agreement as he spoke in a hushed and relaxed tone. They had been married for 40 years. My psychologist colleague and I listened with interest and concern as they discussed their daughter who had severe schizophrenia with depression (schizoaffective disorder), lived on her own and regularly self harmed. Bruce had been living with advanced prostate cancer for which he had received radical surgery some years before and was now on the antiandrogen Casodex. He was happy with the erections produced by intrapenile injections of alprostadil but wanted to try a PDE5is. I discouraged him knowing that the cavernous nerves, which hug the prostate and are needed for PDE5is to work (they work at the nerve vessel junction) had been destroyed at surgery (alprostadil by passes these nerves as it works directly on penile blood vessels). Nevertheless I prescribed tadalfil, a long acting PDE5i. He came back to see me some weeks later and reported that he was happy with the result. This medication did not help the ED but lifted his mood. He and his wife were very grateful. A number of papers, many undertaken in animals, show an uplift in mood, particularly when the mood issue is a low grade chronic issue – termed dysthymia. We reported on this in a short letter in 2009.

In 2006 Daniel Richardson (a junior doctor at the time), myself and colleagues discussed the issue of whether the PDE5Is, in men having sex with men (MSM), promotes the onward transmission of HIV. Logic might suggest that their use in men with ED would result in better erections and hence increased ease of wearing condoms in partnered sexual scenarios. Unfortunately the available evidence at that time suggested that the opposite was true. Studies reported that the concomitant use of recreational drugs such as 'crystal meth' or cocaine, along with the PDE5is, was widespread. These recreational drugs tend to cause ED because they may constrict penile blood vessels. At the same time they cause sexual disinhibition, which may lead to non condom use. We felt it was important that PDE5is are prescribed by healthcare workers who could counsel patients about concomitant use of condoms, ideally without recreational drugs.

A more recent study in gay and bisexual men showed that Truvada – a combination of potent antiretrovirals – is now commonly used as medication to prevent HIV acquisition along with PDE5is and recreational drugs. The addition of Truvada has virtually eliminated the acquisition of HIV during condomless sex.

There are now four PDE5is prescribable in the UK, sildenafil, tadalafil, vardenafil and avanafil.

The patents for the PDE5is have expired except for avanafil. That being so, the generic equivalents are much cheaper. Sildenafil can now be sold over the counter at pharmacies in the UK.

The halcyon days of large scale PDE5i sponsored attendances at the ESSM have therefore long passed, although the ESSM continues to be held annually. In later years there was a greater representation of research involving women, with psychologists attending no longer being a rarity.

https://www.thefreelibrary.com/Meet+Dr+Viagra+...C+my+wife+is+pregnant%2C+but+no%2C+I+don%27t+use+it...-a060654769

Osterloh, IH and Riley, A. Clinical update on sildenafil citrate. British Journal of Clinical Pharmacology. 2002 Mar;53(3):219–23.

Yaman, O, Tokath, Z, Inal, T and Anafarta, K. Effect of sildenafil on nocturnal erections of potent men. International Journal of Impotence Research. 2003 Apr;15(2):117–21.

Goldstein, I, Fisher, WA, Sand, M et al; Vardenafil Study Group. Women's sexual function improves when partners are administered vardenafil for erectile dysfunction: a prospective, randomized, double-blind, placebo-controlled trial. Journal of Sexual Medicine. 2005 Nov;2(6):819–32.

Hames, RK, Wilson, C and Goldmeier D. Oral phosphodiesterase type 5 inhibitors: nonerectogenic beneficial uses – a comment. Journal of Sexual Medicine. 2009 Jul;6(7):2079.

Hammoud, MA, Vaccher, S, Jin, F, et al. The new MTV generation: Using methamphetamine, Truvada™, and Viagra™ to enhance sex and stay safe. International Journal of Drug Policy. 2018 May;55:197–204.

Richardson, D, Goldmeier, D, Bell, C, et al. Do phosphodiesterase 5 inhibitors promote onward transmission of HIV in men who have sex with men? Sexually Transmitted Infections. 2007 Jun;83(3):250.

A regular speaker at the ESSM was the Dutchman Professor Marcel Waldinger who held a chair in sexual medicine in Amsterdam. His English was flawless. He was an international expert in a number of fields of sexual medicine, including premature ejaculation (PE) which I shall discuss below. He was also among the first to recognise that SSRI antidepressants can rarely result in permanent genital anaesthesia, low or no sexual desire, weak or no orgasms and erectile dysfunction – what is now recognised as the post SSRI syndrome. He tragically died not long ago at the early age of 64. I want to devote a few sentences in his memory and to discuss the more unusual aspects of the man.

He was the only son of two Holocaust survivors and attended Jewish primary and secondary schools where at times he was bullied because he did not always conform to his peers' ideas, but rather stood his ground when he felt he was in the right. After medical school he went on to specialise in psychiatry and then sexual medicine. I do not use the terms genius and polymath lightly to describe him. I was in awe of him at ESSM meetings, where he mostly spoke about PE. He listened very carefully to his patients' symptoms and was thus able to discover new syndromes and illnesses. He was one of the most fertile minds in sexual medicine and was able to think 'out of the box'. His lectures on the genetics of PE were, to be quite honest, way above my understanding. As well as being an ideas man, he meticulously recorded his research work in great detail and published over 200 research papers. He was a fierce advocate for evidence-based clinical and scientific research. We sometimes were asked to review the same scientific paper before publication. His long and in-depth criticisms of the submitted research could be devastating in their precision. A paper that seemed reasonable to me was often revealed as totally inadequate by Marcel.

And yet there was another aspect to him. He was a talented artist, painting rural scenes, still life and abstract work in oils and acrylic, as well as a gifted jazz and classical pianist. He had a wicked sense of humour. I found out after his death that he began each day seeing two patients using classical psychoanalysis, a treatment not based in the scientific method.

I wondered how his parents' Holocaust experiences had moulded Marcel. I do not know his parents' story. There is evidence for the intergenerational transmission of Holocaust experiences to the second and third generations, so that offspring are more likely to develop hormone and mental health problems, particularly if both parents were survivors. This may be via the process called epigenetics, where behaviour and environment affects gene expression. Of course it may be via nurture rather than nature, where children are exposed to their parents, who may be wracked with depression, anxiety and PTSD.

I can but imagine that Marcel's parents adapted as best they could to the horrors of concentration camps, while preserving a sense of self and a positive emotional existence. It no doubt took relentless focus and strength to obey the tormentors' commands, and yet find some vestige of meaning in life in that agonising existence. I imagine that Marcel's parents were as single minded as he turned out to be. I have no doubt that he took his parents' survival as a source of great pride.

Dashorst, P, Mooren, TM, Kleber, RJ, de Jong, PJ and Huntjens, RJC. Intergenerational consequences of the Holocaust on offspring mental health: a systematic review of associated factors and mechanisms. European Journal of Psychotraumatology. 2019 Aug 30;10(1):1654065.

Csoka, AB and Shipko, S. Persistent sexual side effects after SSRI discontinuation. Psychotherapy and Psychosomatics. 2006;75:187-8.

Imagine, if you will, a family trip to the cinema. The audience is gripped by a romantic scene. A whispered voice is suddenly heard in the cinema stalls. That well might be me, criticising the portrayal of a sexual scene. Such cinematic representations not infrequently depict an almost perfect male lover, who somehow knows exactly how to induce a long lasting, high level of sexual pleasure in his partner, in spite of never having had intercourse with them before. The couple's physical pleasure, even the first time they get together, just seems to flow effortlessly and seamlessly. The man's technique is flawless and he continues until both partners climax almost simultaneously.

I was, however, silent as I watched *On Chesil Beach*, a film portraying a couple where the woman is seemingly disgusted by the sexual act. Her new husband is woefully ignorant of how he should approach their first time together. At their sexual debut, he fumbles at her clothes, proceeding to ejaculate before vaginal entry. Semen is squirted all over her, even though he is thinking of Harold MacMillan, the then UK Prime Minister! She is disgusted and runs away.

My clinical experience tells me that reality is somewhere between these two extremes.

Many men who do not suffer sexual problems in ongoing sexual relationships, nonetheless can ejaculate rapidly on the first few occasions in a new sexual scenario, with the partner being left upset and disappointed. This is understandable because of the high level of general and sexual excitement, often combined with some concern about the novelty of the situation and the partner's new-to-him response.

What is a normal time to ejaculation and when does it become premature?

Prior to 2008 there were a veritable cornucopia of definitions, including times that ranged from 1 to 7 minutes. In the 1960s the world-renowned American sex therapists Masters and Johnson defined premature ejaculation (PE) in terms of a woman's ability to achieve vaginal orgasms at penile thrusting. This could mean, for example, that if a woman took 30 minutes to orgasm vaginally and her partner orgasmed in 25 minutes, he would be classified as having PE.

In 2005 Marcel Waldinger and colleagues in the UK, Spain, Turkey and the USA recruited 500 couples where the man had been in a stable heterosexual relationship for at least six months, having regular sex and asked them to obtain the average length of time to ejaculation at vaginal sex using a stopwatch. These subjects were drawn from an unbiased selection of each country's population. The median time (not a mean because the distribution was skewed to the right) was 5.4 minutes (range 5 to 44 minutes).

Subsequently the International Society of Sexual Medicine (ISSM) consensus agreement came up with a definition of the time for lifelong PE being 'one minute of intravaginal sex or thereabouts', as well as inability to control the ejaculation, and distress associated with the problem. The average time for the shortest 2.5 per cent of a general population, based on the median time for 'normal' heterosexual men being 5.4 minutes, was also just over a minute. Two further studies of men with premature ejaculation seemed to confirm this time of 1 minute or thereabouts. Looked at another way, almost all men who complain of severe lifelong PE ejaculate within about a minute.

The definition of acquired PE, meaning at some point in the past it has not been rapid, is about 3 minutes.

Some research shows that the prevalence of PE overall in gay men is less than heterosexuals.

Women are affected by PE, not so much because of the short time of intravaginal sex but rather because the men are so focused on their ejaculatory time that neglect the overall sexual needs of their partners. There is some evidence that the aetiologies of primary PE may be due to genetic variations of serotonin, dopamine, androgen and oxytocin in relevant nerve cells in the brain and spinal cord.

Prior to the above genetic data being published, Marcel Waldinger had suggested that PE can run in families. We at St Mary's also felt there might be a genetic component based on our clinical data. In 2005 Dr Daniel Richardson and I had the impression that there was a preponderance of men with PE in our sexual dysfunction clinic at St Mary's who came from Muslim or Asian countries. We therefore undertook a retrospective analysis of the 123 men who presented with PE over an 18 month period, and 60 per cent of these patients came from either Muslim Middle Eastern countries or from an Asian country. The proportions of men from those countries in the local population and our general sexual health clinic were 12 per cent and 10 per cent respectively. Based on this data Daniel undertook an exploratory qualitative study to generate ideas as to why there was a preponderance of PE in men from those backgrounds. He used semi-structured interviews of ten male volunteers who came from an Islamic background with a diagnosis of PE. Interviews were tape recorded and then transcribed. Transcriptions were hand coded and analysed using grounded theory. Themes that surfaced were anxiety about first sexual experience (with subtheme of fear of being discovered having sex and therefore wanting to finish early), sex before marriage, sex outside of marriage, religion, 'stress', exposure to flagrantly sexual Western images, living in the UK and subsequent feeling of sexual freedom. The notion of hurried first acts of sex (e.g. in back seats of cars) had been discussed earlier by Masters and Johnson. In our study some subjects felt that they were breaking religious norms, saying that PE was a punishment from God. They reported that they felt a conflict between living as Muslims and the sexual freedom that they found in the UK. The study results gave us valuable insight into the wider religious and social background of our patients as well as a deeper sense of understanding of our patients' PE.

Secondary PE, whereby there has been a normal time to orgasm in the past, can be caused by prostate inflammation, thyroid issues, relationship problem and post-traumatic stress disorder (PTSD).

Some years earlier I wrote a short paper with Dr Fabian Chen, who now works as a GUM consultant in Reading, discussing PE in evolutionary terms. Most mammals ejaculate very quickly – there is no advantage for prolonged penetration. Chimpanzees are no exception, ejaculating in seconds. This would seem to be an advantage for a non-dominant male so that intercourse can be completed before the female's absence is discovered by the dominant male.

Fabian and I further presented data reported by others of a study where healthy young men aged 20 were masturbated blindfold by a masseuse and told to orgasm when they wished. The average time was two minutes implying that where a degree of control is not necessary (as is the case at solo rather than partnered sex) orgasm can be fast.

Management of primary PE includes the squeeze method, whereby prior to the sensation of orgasm inevitability, the glans penis is firmly pressed till arousal subsides. The stop-start similarly asks the man (or his partner) to stop stimulation and wait for arousal to subside, followed by increasing arousal. In both techniques arousal is diminished and then increased over a number of cycles after which ejaculation is allowed to occur. These behavioural therapies are based on two premises. The first is that the stimulus response is attenuated by gradually exposing the patient to more intense and prolonged sensations just below the threshold for eliciting the response. The second is to familiarise the man with the sensation of ejaculatory inevitability and to recognise the sensations just prior to that event. Recognising pre orgasm inevitability sensations underpins the success of the squeeze and stop-start techniques. It is at this stage that the squeeze and stop-start should be initiated. We successfully incorporated mindfulness into the squeeze and stop-start techniques in a number of our patients. However, to my knowledge there are no published reports of using mindfulness in the current literature of PE. Medications can be very helpful in managing PE. These include local anaesthetic creams put on the frenular area of the penis 20 minutes prior to sex. Short and long acting SSRIs antidepressants work very well but on their cessation the patient is usually back to square one.

PE, or a problem seemingly similar to it, can present in many different ways.

Ali was a 30-year-old man who worked in an Indian restaurant. He had moved from Bangladesh some ten years before, initially being on his own in the UK, and in later years bringing his wife and children to join him. He attended my clinic with his wife. She led the conversation from the outset, cutting to the chase, in effect saying, 'get him fixed so that he doesn't come so fast'. Their history was that the marriage had been arranged. However they got on well with each other. There were no erectile problems. He consistently ejaculated intravaginally in less than 15 seconds. I enquired about foreplay. They both asserted that doing this was not part of their culture. In terms of treatment, I offered psychological support, seeing the couple together with a view to engaging in the stop-start technique and perhaps some mindfulness. She vehemently declined this. She wanted tablets for him. If the outcome of pharmacological treatment was unsuccessful she did not want him back in the marital home. Thankfully he responded very well to sertraline, an SSRI antidepressant. The couple returned after eight weeks. His time to ejaculation was now five minutes. She was happy. They did not want any further intervention so I discharged him back to his GP for him to continue medication.

George was referred to the clinic by his GP. He was a 30-year-old solicitor who had become highly embarrassed over the course of his sex life because he ejaculated rapidly. However this problem was situational and temporally related to the first few times he had intercourse with a new partner. Once he 'got used to' the partner he could continue intercourse for 20 minutes or thereabouts. The first few times the ejaculation could be so severe as to be 'ante portas' (Latin for 'in front of the door'). He himself had come up with a solution of sorts – that was to ejaculate by solo masturbation before seeing the partner. However he was not entirely happy with that as an answer. We discussed the treatment options. He tried putting some local anaesthetic cream on the frenulum area of his penis 20 minutes prior to sex. This seemed to work for him. I am not sure if it was a placebo effect.

Keiran, who was 45 years old, had had a good sex life with his partner Damian for ten years. He then developed pain in perineum, that is behind the testes as well as some discomfort passing urine. After investigation in our department he was found to have prostatitis, that is inflammation of the prostate gland. No cause for this could be detected. He did not respond to a course of antibiotics. At the same time he complained of an inability to sustain an erection. His GP had given him sildenafil but it did not help. When I saw him he told me his penis just became soft soon after starting sex. On closer enquiry it was apparent that this was PE rather than an erection problem. He was treated accordingly. Patients who complain that their erection fades at sex may be actually describing either PE or erectile dysfunction or both. Careful questioning will reveal exactly what the problem is!

Abbas was a 40-year-old man who had been repeatedly tortured in a Middle Eastern country by a radical Muslim terrorist group. Each day he heard gunshots and men screaming in adjacent cells and told in no uncertain terms that his turn would be next. After months in jail he was released only for the country of his residence to accuse him of collaboration with the terrorists. He was transferred to another country for trial. In mid flight he brandished a knife, threatening to use it unless the plane was rerouted. He ended up at Stansted airport and asked for asylum there. He was held in remand for months at Belmarsh prison which he told me was worse in many ways than his time spent in prison in the Middle East. His family were allowed to come to the UK as refugees. It took many court appearances over 14 years for him to eventually gain UK citizenship. He always told me it was never his intention to harm anyone with the knife – he was just desperate. The British justice system finally concurred. In many ways he was very resilient and showed only two signs of post-traumatic stress disorder (PTSD). One was occasional nightmares and the other was very rapid ejaculation – his intravaginal time was about 30 seconds whereas pre torture it had been about 10 minutes. PTSD is an important cause of secondary PE and easy to overlook. Abbas's nightmares responded to EMDR (eye movement desensitisation and reprocessing). However, the PE did not resolve, but an SSRI fluoxetine increased the time to five minutes. He happily remains on this medication five years later. The mechanism of why PE occurs in PTSD likely relates to over stimulation of the amygdala area in the brain via upregulated adrenergic neurones.

Anand was a 25-year-old man who had come to the UK from India. He had no sexual partner. However whenever he stood next to women at work he would 'leak semen'. This also happened at the end of micturition. He did not have a sexual partner nor did he masturbate. He felt this problem was PE. Each time he produced semen or what he thought was semen he felt 'weak'. I explained that he had the 'Dhat' syndrome, which is a culture bound complex of preoccupation with loss of pre cum or overflow from the prostate gland produced at the end of micturition, particularly when genital fluids build up because of sexual abstinence. In Indian culture, or at least some parts of it, 'one ounce of semen is the equivalent to 100 ounces of blood' in terms of bodily energy. After a number of sessions and much reassurance and explanation the symptoms resolved, possibly because he had found himself a girlfriend.

Bert came to my department complaining that he ejaculated after a minute of thrusting. His girlfriend was accepting of this as they had highly enjoyable foreplay. However Bert, who was a 39-year-old disc jockey, was determined not to be beaten by his fast time at vaginal sex. He had not found the stop-start technique useful. He declined any pharmacotherapy. So my psychologist colleague and I invited him to try mindfulness. In essence he learned how to be in any moment – but just that moment and not to be drawn into catastrophic or future thinking. Most people find they need to work on mindfulness daily for some time before results are apparent. But not Bert. After a few weeks of learning the basics of mindfulness he was able to just be present in arousal and attend to the sexual sensations while noticing his thoughts and fears of ejaculating too soon rather than becoming involved in them. He later told me he could sustain penetration for up to ten minutes.

Goldmeier, D and Chen, F. Premature ejaculation. Is it a biogenic or psychogenic disorder? An ethno-evolutionary rejoinder. Sexual and Marital Therapy. 1999;14:91–2.

Richardson, D, Wood, K and Goldmeier, D. A qualitative pilot study of Islamic men with lifelong premature (rapid) ejaculation. Journal of Sexual Medicine. 2006 Mar;3(2):337–43.

Richardson, D and Goldmeier, D. Premature ejaculation – does country of origin tell us anything about etiology? Journal of Sexual Medicine. 2005 Jul;2(4):508–12.

Richardson, D, Goldmeier, D, Green, J, Lamba, H and Harris JR; BASHH Special Interest Group for Sexual Dysfunction. Recommendations for the management of premature ejaculation: BASHH Special Interest Group for Sexual Dysfunction. International Journal of STD & AIDS. 2006 Jan;17(1):1–6.

Mostafa, T, Abdel-Hamid, IA, Taymour, M and Ali OI. Gene Variants in Premature Ejaculation: Systematic Review and Future Directions. Sexual Medicine Reviews. 2020 Oct;8(4):586–602.

Waldinger, MD, Quinn, P, Dilleen, M, Mundayat, R, Schweitzer, DH and Boolell, M. A multinational population survey of intravaginal ejaculation latency time. Journal of Sexual Medicine. 2005 Jul;2(4):492–7.

Serefoglu, EC, McMahon, CG, Waldinger, MD et al. An evidence-based unified definition of lifelong and acquired premature ejaculation: report of the second International Society for Sexual Medicine Ad Hoc Committee for the Definition of Premature Ejaculation. Journal of Sexual Medicine. 2014 Jun;11(6):1423–41.

Barbonetti, A, D'Andrea, S, Cavallo, F, Martorella, A, Francavilla, S and Francavilla, F. Erectile Dysfunction and Premature Ejaculation in Homosexual and Heterosexual Men: A Systematic Review and Meta-Analysis of Comparative Studies. Journal of Sexual Medicine. 2019 May;16(5):624–32.

Ventriglio, A, Ayonrinde, O and Bhugra, D. Relevance of culture-bound syndromes in the 21st century. Psychiatry and Clinical Neurosciences. 2016 Jan;70(1):3–6.

Riley, A and Segraves, RT. Treatment of premature ejaculation. International Journal of Clinical Practice. 2006 Jun;60(6):694–7.

Masters, WH and Johnson, VE. Human Sexual Response. Toronto; New York: Bantam Books, 1966.

Chapter 17

THE INTERNATIONAL SOCIETY FOR THE STUDY OF WOMEN'S SEXUAL HEALTH (ISSWSH), SEXUAL DESIRE AND AROUSAL IN WOMEN

In late October 2001 I flew to Boston USA with my research nurse colleague Miranda Cowen. The city presented itself as exuberant and outgoing. The autumnal trees at the banks of the Charles River were proudly flaunting their yellow, brown and red leaves. The chatter at the hotel desk extolled the talents and strengths of the Boston Red Sox, the oldest baseball club in the country. Waiters at restaurants unashamedly asked for 30 per cent gratuity over and above the billed amount. We were attending a meeting of the Female Sexual Function Forum (FSFF). The powerhouses behind the FSFF were the Goldsteins. Irwin Goldstein at that time held a chair in Sexual Medicine at Boston University's academic department of Urology. He spoke in encouraging superlatives, describing topics that interested him as 'awesome' in loud, animated and passionate fashion. He was as much as scientist as a compassionate physician, having been a biomedical engineer prior to studying medicine at McGill University. Irwin's wife Sue with her background in music, technical writing and teaching was also a bundle of enthusiastic energy. Whenever I saw them together, they were happily engaged in each other's company. Their combined energy was greater than the sum of their individual fervour, if that seemed at all possible! At the meeting of the members of FSFF the Goldsteins, more than any other individuals, pushed for the society to change its name. And so ISSWSH was born. Sandra Leiblum was its first president. In the early years it was truly an international society with a multidisciplinary and thought-provoking approach to women's sexual health.

In the early Spring of 2004 ISSWSH held its annual meeting in Atlanta Georgia. The city is famous for the Battle of Atlanta in 1864 in the American Civil War, where the Union overcame the Confederate army. There is a museum that commemorates the event. I visited that as well as the house of Margaret Mitchell the author of *Gone With the Wind.*

One of the topics at the meeting was 'female ejaculation' at orgasm.

I had previously seen a 45-year-old woman who complained that she soaked the bed at orgasm. This had only been happening over the previous few years. It prevented her from making a sexual relationship, fearing a critical response from any prospective partner. She was well, and on no medication. I assumed that the fluid she complained of was urine, although I had seen case reports of women producing smaller volumes of what looked like semen at ejaculation. I did not have the facility or time to investigate the fluid in detail. Instead I ordered a bladder ultrasound, to be undertaken after she felt she had emptied her bladder. The bladder still contained about 200 ml of urine. I therefore presumed this was being expelled at orgasm. I referred her to my physiotherapy colleague who helped her optimise bladder control and emptying. After that she no longer had a problem and swiftly found a partner.

In 1983 Beverley Whipple, who is currently a Professor Emerita at Rutgers University, wrote a paper on female ejaculation and the G spot, or to give it its full name the Gräfenberg spot, named after the German American physician of that name. She and her co-authors reported on four women who were examined by two gynaecologists who confirmed an area in the front wall of the vagina that when stimulated led to orgasm and ejaculation. Examination of the ejaculate found it had the biochemical characteristics of urine. Some of the non ejaculators also were found to have G spots. A much more recent study from France confirmed bladder emptying in women who 'squirted' at orgasm, along with biochemical analysis of the ejaculate being consistent with that of urine.

In contrast, a study in 2007 reported two premenopausal women who ejaculated during orgasm who were found to have ultrasound images consistent with glandular tissue around the urethra. Direct inspection of the area using a urethroscope confirmed glandular tissue which opened into the urethra. The biochemical analysis of the ejaculate had the features of prostatic fluid.

A 2019 review of the literature concerning the G spot concluded that it may consist of engorged blood vessels but is not responsible for ejaculation. Seemingly the greater the thickness of the space between the front wall of the vagina and the urethra the more likely a woman is to have a vaginal orgasm. My own view is that the G spot may possibly be the body of the clitoris (not the glans which is the part visible in the vulva) which runs in front of the vagina and is engorged at arousal.

To return to Atlanta. A delegate at the conference, who had previously worked as an actor making sexual movies and was now a sex therapist, told a few of us that she was able to 'squirt' at orgasm. Wanting to prove this to us and not being in the least self-conscious, she invited about ten of us to view her 'squirting' in her hotel room. In a very matter of fact way she told us, by way of a running commentary, that she was now milking the G spot by digitally running two fingers on the front vaginal wall from higher to lower repeatedly. She did not appear to orgasm, but told us when the fluid discharge was imminent. At this point Beverley Whipple produced a wine glass and caught the ejaculate, which certainly looked like semen. Beverley announced to the gathering that it had the colour, consistency and odour of semen. My psychologist colleague Agnes was in the room. I looked at her immediately after the ejaculation. Her face seemed to portray a mixture of disgust and fascination. In spite of this she later told me she found the event very informative.

The 2005 ISSWSH meeting was held in Las Vegas. The Goldsteins' presence loomed large at this meeting as usual. Irwin's concern for his fellow man and scientific expertise combined as he saw me riding an expensive racing bike up and down the 'Strip' in Las Vegas at the 2005 ISSWSH meeting. 'You'll get erection problems if you do that David' was his refrain. In fact my genitals were working fine. Irwin was to publish a cautionary article in the *Journal of Sexual Medicine* in 2010 showing that the saddle pressure of racing bikes can dig into the perineum and so cause erectile problems and genital anaesthesia in men and women. Leaning forward and having a down tilted saddle 'nose' helps to prevent such issues.

Some of the Goldstein's energy and enthusiasm rubbed off on me, so that I took on the role of ISSWSH Treasurer in 2008. My attempts at urging physicians with high incomes to part with charitable money to bolster the ISSWSH coffers was abysmal. At the board of directors meeting I spluttered out the basics of the Society's break-even year. Pharmaceutical sponsoring and subsidising meetings was not as easy as the early days of the ESSM, as there were no equivalent 'big' products such as Viagra for women.

The 2009 meeting was held in Florence, Italy. The local scientific committee had arranged for the Board of Directors to have unusual sleeping accommodation – a converted monastery where each of our bedrooms was a somewhat barren monk's cell. Alan Altman, a prestigious US gynaecologist on the board, would not countenance sleeping in such ascetic surroundings and so checked into the local Hilton. ISSWSH at first carried out its accepted policy of holding one meeting out of North America every three years, but later on could not adhere to this policy stating that 'out of USA' meetings were always a financial disaster. In that sense it cannot really be described as an 'International' Society.

Rates of female sexual dysfunction (FSD), sexual desire and arousal were often highlights of ISSWSH meetings and were hotly debated. A milestone in the understanding, or rather misunderstanding of female sexual dysfunction (FSD) was a paper published in 1999 stating that the prevalence of sexual dysfunction in women between the ages of 18 and 59 in the USA was nearly 43 per cent. This much quoted population study was led by a prestigious sociologist Edward Laumann, at the University of Chicago. Sandra Leiblum felt that real dysfunction was not nearly so common, and John Bancroft, then the Director of the Kinsey Institute, felt that much of what the women in that study where complaining of was not a dysfunction but rather related to complaints of stress, tiredness and poor relationships, and not a medical problem per se. This figure of nearly half the women in the USA having a sexual problem was often quoted at early ISSWSH meetings. Pfizer sponsored the 2002 ISSWSH meeting on the basis of work showing that Viagra engorged female genitalia. Ray Moynihan, a lay medical author, wrote a highly critical article in the *British Medical Journal* in 2003 suggesting that FSD had been over medicalised, and that the drug companies were colluding in presenting FSD as a purely medical condition. Later work undertaken by the National Survey of Sexual Attitudes and Lifestyles (NATSAL) in the UK, another large unbiased study, reported sexual complaints in women at 51 per cent, but when combined with the presence of distress, the figure fell to 11 per cent, which can be understood to be an estimate of meaningful and non-transient FSD.

About 30 per cent of women admit to having low sexual desire as a complaint. Those who are additionally distressed by the problem amount to about 10 per cent of any population – quite a sizeable amount. Around 8 per cent of women complain they have problems becoming sexually aroused. But what exactly are desire and arousal? How do they differ?

In simplistic terms, sexual desire can be compared to hunger for food, both involving urges that require action to achieve satiety. Sexual desire however, imperceptibly merges into sexual arousal, which can be understood as being sexually excited or 'turned on'. There is therefore great overlap between these two states, which have been somewhat artificially dichotomised in an attempt to obtain greater understanding of sexual responses in women. Sexual arousal in women has two components. Subjective arousal is the excitement that the woman feels both in the genital area as well as a more generalised pleasurable feeling appreciated in many other parts of the body. Genital arousal consists of the objective engorgement of the labia, clitoris and vagina along with vaginal lubrication. Vaginal lubrication is not a secretion from glands but a seepage through the vaginal wall (called a transudate) from the blood vessels beneath the vagina, which engorge at sexual arousal. In men there is a good correlation between subjective and genital arousal. These parameters can be measured in a laboratory setting, so that if a man sees an erotic film and becomes erect, he is also likely to feel sexually aroused and excited. There is a much poorer correlation between subjective and genital arousal in women. Many women complain that they do not feel sexual excitement but in fact can be shown to genitally engorge efficiently.

The notion of sexual desire itself is also not straightforward. A woman might think about her lover, or see her/him and become desirous to have sex. This is termed spontaneous desire and was first described by Helen Singer Kaplan in New York more than 40 years ago. It is more likely to occur in a new relationship and at certain points in a woman's menstrual cycle (the middle or towards the end). However there are many women who do not have spontaneous desire but rather start from a sexually more neutral place. These women, with so-called responsive desire, have reasons to be willing to engage in sex and have motivation to enter a sexual scenario. These may be the anticipation of physical or emotional closeness of being with the partner or expectation or anticipation of sexual pleasure. Once engaged with her partner she picks up physical cues, for example genital wetness, becomes subjectively aroused which leads on to the responsive desire. Responsive desire was first discussed by Rosemary Basson, an Oxford medical graduate who spent most of her working life in Vancouver, Canada. Of course responsive desire was present in plain sight for all to see, but it took Rosemary's observation, inspiration and courage to define and describe it in 2000. Rosemary was at pains to point out that responsive desire was a normal variant, and should not be pathologised. That said, many women (and indeed some men) who consulted me over the years with what amounted to responsive desire complained bitterly of it. They knew that if they went with a new partner that they would once again revert to spontaneous desire, but did not wish to end their current relationship. It was a real problem and certainly felt like an abnormal response to them!

To complicate matters further, some experts have used the term sexual 'drive' rather than desire to describe an urge to have sex that is not related to a person or situation but is rather innate, much as thirst does not need to be a response to seeing a pleasant drink.

A number of medications have been trialled for treating low desire, using the definition of hypoactive sexual desire disorder (HSDD). HSDD is defined as low sexual desire with distress lasting six months or more.

As the new millennium passed, I was approached by the German pharmaceutical firm Boehringer Ingelheim who had a large UK subsidiary operation in Bracknell Berkshire. The parent company in Ingelheim in Germany has a large research and production site near the River Rhine which is about 60 km west of Frankfurt. It was founded at the end of the 19th century and is still owned privately by the Boehringer family. The company had passed through difficult times in World War 2 when its pharmaceutical manufacturing plant had been staffed by foreign slave labour. Boehringer had undertaken trials of flibanserin which was intended as a serotonin altering antidepressant. However it did not show efficacy in that regard, but did surprisingly demonstrate some increase in sexual desire in the subjects. That being the case, the company now wanted to undertake trials of the medication in women in Europe and North America. I was chosen as principal investigator for the UK sites. My initial task was to lecture to the Bracknell team on the topic of low sexual desire in women. Later on, in 2007 I presented the detailed ethics application to a committee in Oxford, who I think had never had to assess a study involving sex and were just a little shy when questioning me.

In order to show efficacy the trial needed a valid method to measure sexual desire. The American Foods and Drugs Administration (FDA) at that time deemed the best measure was to quantify how many 'satisfying sexual events' had taken place per unit time, as well as to ask how sexually desirous a subject felt at various points in the day. Relying on filling in paper diaries was known to be notoriously inaccurate, in large part because subjects filled them in retrospectively at a time remote from the event. Boehringer therefore used a novel approach at the time – giving subjects hand held electronic diaries which prompted them a few times a day, asking, 'Did you have desire to have sex since your last diary entry?', and to rate the desire 'low, medium or high'. If sex took place the subjects needed to give an indication as to how satisfying the event was. This was a few years before the launch of the first Android and iPhones, which today, no doubt, would be used to electronically prompt subjects. The handheld electronic diaries were tried out in small pre study groups of women who had no sex problems. One of these volunteers was a 50-year-old Imperial College employee who had responded to our local advert for volunteers. She did not consider herself to have sexual problems. She felt that the electronic diary prompting was really intrusive and embarrassing. Her teenage daughter asked what the bleeping noise was, as the e-diary loudly and frequently prompted her. After making excuses and 'telling a porky', she sat down, feeling exhausted after a days' work at Imperial College, and now cooking for the family and helping with homework, thinking 'sex is about the last thing on my mind'. The outcome of the various studies showed that, taking flibanserin daily, on average increased the number of sexually satisfying events per month by 0.5 compared to placebo. Flibanserin is no Viagra in terms of efficacy. Its prescribing in the USA is restricted, in part because of its side effect profile. It does not have a UK license.

After much criticism by experts, most of whom were the great and good at ISSWSH, the FDA agreed to change the primary measure of desire and arousal from satisfying sexual events to the use of well validated questionnaires.

Flibanserin was much discussed at ISSWSH over the years.

Siobhan was a 38-year-old shop assistant who was in a seven-year relationship with Charles. When they met she was very turned on, and thought about him all day long. At sex she became excited and could orgasm easily both vaginally and clitorally. After a year, during which time she had moved in with him, she no longer had spontaneous sexual thoughts about him. However they got on very well and she was emotionally close to him. That being so, she was happy to just go with his sexual advances, as she enjoyed the closeness it afforded. After a few minutes foreplay she felt sexually excited and was keen to take sex further, being able to orgasm. This scenario of having spontaneous desire for a few months that later turned to 'responsive' desire had happened in her two previous relationships. In spite of being told that her sexual response, i.e. responsive desire, was a normal variant, she continued to complain about having no spontaneous desire, except when talking to a work friend who was keen to date her. She had tried a number of medications all of which had not helped. We lost contact with her after a few months.

Rosemary was a 40-year-old premenopausal woman who had suffered from lack of sexual desire and arousal since she was sexually assaulted five years previously. Her partner Vivienne had been very supportive all along. There had been no flashbacks to the event and no other psychiatric symptoms. Vivienne had noticed when she tried to arouse Rosemary that there was as much wetness as pre sexual assault. When seen in clinic they both almost demanded some Viagra. We explained to the couple that because of the assault there was a brain disconnect between subjective arousal and lubrication. She tried Viagra a few times and indeed it did increase the amount of wetness at arousal but did not affect the lack of feeling sexy. After trauma therapy (EMDR) she made an excellent recovery.

Charlene was a 42-year-old van married driver who had become depressed a year prior to attending the clinic. She said she felt very little pleasure nowadays from sex, whereas before her depression she described a successful and enjoyable sex life. She was no longer depressed, but remained on an SSRI antidepressant. While waiting for psychological treatment we prescribed a small dose of sildenafil (Viagra) to take an hour before sex. Although her sexual desire was still a problem, at stimulation she responded well to her partner's advances and enjoyed arousal and could orgasm clitorally albeit with some effort on her husband's behalf. Sildenafil may help women on SSRI antidepressants to regain subjective arousal.

Sommer, F, Goldstein, I and Korda, JB. Bicycle riding and erectile dysfunction: a review. Journal of Sexual Medicine. 2010 Jul;7(7):2346–58.

Moynihan, R. The making of a disease: female sexual dysfunction. British Medical Journal. 2003 Jan 4;326(7379):45–7.

Mitchell, KR, Mercer, CH, Ploubidis, GB et al. Sexual function in Britain: findings from the third National Survey of Sexual Attitudes and Lifestyles (Natsal-3). The Lancet. 2013 Nov 30;382(9907):1817–29.

Stone, AA, Shiffman, S, Schwartz, JE, Broderick, JE and Hufford, MR. Patient non-compliance with paper diaries. British Medical Journal. 2002 May 18;324(7347):1193–4.

Kingsberg, SA and Althof, SE. Satisfying sexual events as outcome measures in clinical trial of female sexual dysfunction. Journal of Sexual Medicine. 2011 Dec;8(12):3262–70.

Nurnberg, HG, Lauriello, J, Hensley, PL et al. Sildenafil for sexual dysfunction in women taking antidepressants. American Journal of Psychiatry. 1999 Oct;156(10):1664.

Parish, SJ, Cottler-Casanova, S, Clayton, AH et al. The Evolution of the Female Sexual Disorder/Dysfunction Definitions, Nomenclature, and Classifications: A Review of DSM, ICSM, ISSWSH, and ICD. Sexual Medicine Reviews. 2021 Jan;9(1):36–56.

Basson, R. The female sexual response: a different model. Journal of Sex & Marital Therapy. 2000 Jan–Mar;26(1):51–65.

Gravina, GL, Brandetti, F and Martini, P. Measurement of the thickness of the urethrovaginal space in women with or without vaginal orgasm. Journal of Sexual Medicine. 2008 Mar;5(3):610–8.

Ostrzenski, A. G-Spot Anatomy and its Clinical Significance: A Systematic Review. Clinical Anatomy. 2019 Nov;32(8):1094–101.

Wimpissinger, F, Stifter, K, Grin, W and Stackl, W. The female prostate revisited: perineal ultrasound and biochemical studies of female ejaculate. Journal of Sexual Medicines. 2007 Sep;4(5):1388–93.

Salama, S, Boitrelle, F and Gauquelin, A. Nature and origin of 'squirting' in female sexuality. Journal of Sexual Medicines. 2015 Mar;12(3):661–6.

Goldberg, DC, Whipple, B, Fishkin, RE, Waxman, H, Fink, PJ and Weisberg, M. The Grafenberg spot.

Chapter 18

MORE ON STAMMERING

With the new millennium, freedom from opiates and a growing national and international reputation in sexual medicine, came offers to lecture. I was in a dilemma of my own making. Colleagues wanted to hear what I had to say. GPs sought updates on Viagra and its sister drugs. Trainee GPs wanted broad talks on sexual dysfunction. GU physicians wanted to hear about sexual dysfunctions as they related to GU medicine and HIV. I was considered an expert on PGAD and another rare syndrome of post orgasmic illness (POIS see below). I guess I became a little puffed up with pride in what I had achieved. But having some self-confidence was no bad thing. And yet ... and yet, my stammer still haunted and hounded me. Sometimes there is humour to be found in the face of suffering. A woman patient with a stammer stormed out of my consulting room, screaming at the receptionist that I had impersonated her stammer. She was told 'Dr Goldmeier's stammer can be bad sometimes'. She contritely re-entered my room and told me what had transpired. We shared a brief empathic moment and continued with the consultation.

I prepared my talks meticulously. I knew in advance what words or phrases would be difficult to enunciate. Dr Spencer Brown of Ohio University had researched what parts of speech were most difficult for stammerers over 70 years ago. Nouns, verbs, adverbs and adjectives are more difficult to enunciate than articles (the, a), conjunctions and pronouns. Long words were harder than short words and words at the start of a sentence harder than those within a sentence. Finally the initial phonic or sound was more difficult if a consonant rather than a vowel. Later research confirmed that the stammerer anticipates which words she or he will have trouble even before enunciation begins. And as the years passed

I came to an understanding that I probably suffered from a degree of inattentive rather than hyperactive ADHD. I could make careless errors, be inattentive if bored, distracted by noise and would often feel an urge to get up and leave meetings. In order not to let these issues become problematic with patients I had over the years become very careful to check my interactions with patients and I believe in over 50 years of practice I never made an error because of inattention. I was helped by having a moderately high level of anxiety, supplemented by liberal amounts of caffeine before a lecture. This combination was my de facto 'psychostimulant' rather than the conventional amphetamine like drugs, which would have both raised my blood pressure to danger levels and be unsuitable for an addict to take, as I would crave more and more as time went on. On completion of the lectures I would often feel utterly exhausted and drained. I would then revert to having problems thinking and structuring what I wanted to say as well as producing the final utterances in speech. That being the case I could swing from almost complete fluency to a complete verbal block depending on energy levels. My ADHD over the years did confer at least one advantage – creativity. There is reasonable evidence in the literature to confirm my suspicions. I feel I had the edge on my colleagues in terms of coming up with new ideas for clinical research. Many of my fluent speaking colleagues could give long, unprepared, perfectly executed monologues but did not have the gift of having new ideas. My creativity also found expression in poetry, drawing and painting.

I rarely relished the idea of standing on a podium, with screen to one side and hundreds of delegates in front of me, expecting to be informed and taught – not pained by having to listen to 'a blabbering fool'. My comments to myself as I strode to the podium might be:

'I'm the expert, but I can't get the words out, I'm stupid, I'm a sluggish thinker, I'm going to embarrass myself and the delegates, I wish I could disappear, I'm angry with the world for making me stammer. People don't understand about stammering. I don't understand my stammer and why can't I be a fluent speaker'.

In rare moments of insight I would congratulate myself saying that in spite of being a stammerer, I had achieved a great deal.

I enjoyed preparing PowerPoint slides to go with the talks, illustrating them meticulously with imported photos and videos. The projected slide data helped, as the words that I felt I would have trouble with could be highlighted by my laser beam pointer. That made me feel that the audience already knew the word or phrase, so that voicing it was merely a supplementary gesture.

Thankfully, in the 2020s, stammering has become more acceptable to society. That said, there are still stories of stammerers not being able to find employment after umpteen interviews and committing suicide. On the positive side, Wikipedia has a list of 114 well known stammerers, including politicians, musicians, writers and actors. The singer Megan Washington bravely gave a TEDx talk about her stammer.

The immediate cause of the stammering speech defect is of course based in brain pathology. However, Dr Christopher Constantino, who himself is a stammerer, has pointed out that society and the listener play their part in the end product of the stammerer's distress. The social model of stammering has little to do with dysfluency but everything to do with socially stigmatised speech that is not accepted by society. The stammerer is therefore punished by society for violating the concept of able-bodiedness. Stammerers internalise this concept and put blame on themselves. Constantino further points out how society fails to understand situations such as phone calls which can be so difficult for the stammerer. I have had the phone put down on me many times. I am struggling to enunciate the words. The listener hears silence in response to their question and so terminates the call. The last time I went to vote at a polling station, the local officer asked my name. I stood in front of him trying to push out my name. The officious twit concluded that as I didn't know my name I was a fraudulent voter. He told me to leave the polling station. I was fuming and at the same time so ashamed. I have since voted by post. Interviews, as stated, can be a nightmare, as stammerers may circumlocute and therefore not say what they mean, thus producing a poorer word content reply. The social model of stammering removes the blame and responsibility of stammering from the sufferer, who is left just with a speech defect.

Brown, SF. 1945 The loci of stuttering in the speech sequence. J Speech Defects. 10:181–92.

Constantino, C, Campbell, P and Simpson, S. Stuttering and the social model. Journal of Communication Disorder. 2022 Mar–Apr;96:106200.

Tichenor, SE, Johnson, CA and Yaruss, JS. A Preliminary Investigation of Attention-Deficit/Hyperactivity Disorder Characteristics in Adults Who Stutter. Journal of Speech, Language, and Hearing Research. 2021 Mar 17;64(3):839–53.

Jackson, ES, Yaruss, JS, Quesal, RW, Terranova, V and Whalen, DH. Responses of adults who stutter to the anticipation of stuttering. Journal of Fluency Disorders. 2015 Sep;45:38–51.

Hoogman, M, Stolte, M, Baas, M and Kroesbergen, E. Creativity and ADHD: A review of behavioral studies, the effect of psychostimulants and neural underpinnings. Neuroscience & Biobehavioral Reviews. 2020 Dec;119:66–85.

Girard-Joyal, O and Gauthier, B. Creativity in the Predominantly Inattentive and Combined Presentations of ADHD in Adults. Journal of Attention Disorders. 2022 Jul;26(9):1187–98.

Chapter 19

THE CHANGING FACE OF HIV

Until the mid 1990s, the death rate from HIV remained high in spite of antiretroviral (ARV) drugs inhibiting reproduction of the virus. HIV reproduces rapidly but has no enzyme to correct mistakes in its genetic code at reproduction. These errors appear as mutations and can therefore render any one ARV useless. This meant patients would again become ill with HIV once the virus that inhabited their bodies had mutated. In 1996 combined antiretroviral medication was initiated in Western countries. By combining two or more antiretrovirals, the chances of HIV mutations causing treatment failure were greatly minimised. At St Mary's this meant that the HIV wards gradually emptied. HIV was changing from a lethal disease to a long-term chronic illness. Patients on this new combined treatment could now live an almost normal lifespan. More than one of my colleagues told me they missed the excitement of the life and death struggle, which had now been largely superseded by the more banal management of the side effects of the combined ARV therapy. By 2010 patients with newly diagnosed HIV were started on lifelong therapy as soon as the diagnosis was made, whatever the stage of their disease. The world now had space to look back at the devastation that HIV had wrought. As of 2022, the HIV epidemic worldwide has claimed about 36 million lives compared to a total of 8 million for COVID-19. There are almost endless lists on the web of those who had succumbed to HIV. They included Freddie Mercury who would now be in his late 70s, Rudolf Nureyev who would now be in his late 80s and Isaac Asimov the scientist and science fiction writer who was infected by a blood transfusion after cardiac surgery. He would now be over 100.

All that said, the side effects of the new ARV therapy can be troublesome, including sexual problems.

Charles was a 45-year-old gay man who developed AIDS in 1996, just as highly active antiretroviral treatment (HAART), another term to describe combined ARV therapy for HIV, was introduced. He was a tall man of few words, mutton chop sideburns and a weathered facial complexion. His treatment included lamivudine as well as other antiretrovirals. Lamivudine acts by inhibiting the conversion of virus RNA to DNA. Although his HIV disease improved, he developed numbness of his hands and feet and at the same time suffered erectile dysfunction (ED) and an inability to ejaculate. These issues were caused by lamivudine associated nerve damage – a neuropathy. He was initially tried on sildenafil (Viagra) for the ED. However this was unhelpful, likely because nerves going to the penis needed to be intact for sildenafil to be effective. He was therefore taught how to give injections of alprostadil into his penis, which rapidly produced good erections. Alprostadil works directly on penis blood vessels, so that his impaired nerves did not matter. A majority of patients with ED opt for tablets rather than injections because of the convenience of oral therapy. However compared to tablets, injections have a lower failure rate and produce erections more rapidly and without the need for sexual stimulation. Charles's delayed ejaculation however persisted and was likely also due to the neuropathy – in particular dysfunctional small nerve fibres failing to send sensory information from the penis to the spinal cord and brain. Other causes of a neuropathy were excluded. Because of this sexual problem he stopped taking his antiretroviral medication a number of times, resulting in HIV illness relapse. The ability to ejaculate improved somewhat each time he stopped treatment. As the years went by, alternative HIV medication was prescribed that enabled him to stop the lamivudine. He continues to find ejaculation troublesome, but is able to do so with the appropriate sexual stimulation and fantasy.

The side effects of HIV and its medication can be very visually apparent, and lead to much distress. Lipodystrophy, that is, a condition where bodily fat is altered and redistributed seems to be caused by HIV itself as well as some ARVs.

Grace was a 35-year-old woman who came to the UK from Zimbabwe in 2000. She had managed to obtain treatment for her HIV there with zidovidine – the original ARV, not used much in the UK in the new millennium. However by the time Grace saw us she had ceased taking it, because she felt it had altered her body shape. When first seen in our unit, her HIV had rendered her immune system almost absent, in that her CD4 count was 20. Her arms and legs had become thinner and her abdomen had enlarged because of fatty deposits. She had also developed a pronounced fat pad under her chin, while her cheeks had become sunken, revealing the bony structure beneath. In Zimbabwe her altered bodily features had earmarked and stigmatised her as having HIV. Her friends no longer wanted to know her. She had been successful in her application to come to the UK as a refugee. We started her on an up-to-date HAART regime. She responded well over a few month in terms of her immune system. However her bodily shape became more accentuated and extreme. Although she wanted to have a relationship, she felt unable to even contemplate this while her lipodystrophy persisted.

She was given an exercise regime, had liposuction of the accentuated areas and had facial fillers injected into the sunken areas. Although her physical features were normalised she still complained of low sexual desire. Given more time, she divulged how depressed she was living away from her home country, and how socially isolated she had become in the UK. Antidepressants and cognitive therapy helped her depression although her low sexual desire persisted.

In a review article written by myself with colleagues in 2005, sexual dysfunction in women in Sub-Saharan Africa (SSA) was discussed. We documented widespread male coercion and violence in sexual relationships in SSA at that time. Poverty, lack of access to HIV treatment, fear of disclosure (could result in violence, job loss, homelessness and even murder) were issues that women from these areas had to suffer. Not surprisingly this resulted in low sexual desire in these women, as well as inability to discuss sex and sexual problems. Sexual dysfunction in women with HIV can have physical causes, including hormone abnormalities. However, psychosocial issues are just as important. Lipodystrophy can further stigmatise these women.

Our unit saw many women from Sub-Saharan Africa with HIV who described the profound and fearful psychosocial issues they had to encounter.

In 1999 I was asked to see Patience, who was on our wards at St Mary's. She was a 29-year-old Black woman from Cape Town visiting her mother who had moved to the UK some years before. Her long sleeved, loose fitting, multicoloured dress almost completely covered her intravenous port line. She answered my questions in accented but perfect English. Her demeanour reflected her name. Patience suffered from headaches and some impairment of vision, and she had been found to have advanced neurosyphilis and was now having treatment for that with intravenous penicillin. While in the ward her HIV test had come back as positive. Her CD4 count was very low at 40. I was giving an opinion on her syphilis treatment which seemed to be progressing well. I was interested in how Patience thought she might have acquired HIV – and for that matter syphilis. She thought it likely her husband, who was still in South Africa (SA), had given it to her. As a psychology student she had a keen interest in the social anthropology of HIV in SA and told me that, in spite of Mandela taking over from Botha as president, the rates of HIV were fast approaching 20 per cent of all the population. AIDS and HIV in some sections of the Black population were not directly spoken about, rather referred to as 'pneumonia' or 'TB'. Using these terms was not a 'denial', but rather they were euphemisms used to maintain hope of recovery and to avoid 'disrespect' entailed in referring directly to the nature of HIV stigma. The poor, uneducated and those lacking in literacy were the most prone to acquiring HIV, and the cause of AIDS, when spoken of was not infrequently blamed on witchcraft or religious punishment. Poverty forces many women to earn money by sex work, and even then they are harassed by the police who would want bribes or sexual favours in exchange for non arrest. I am pleased to say Patience made a complete recovery from her syphilis infection and her HIV is under good control.

At about the same as I saw Patience, my colleagues and I did a survey of patients who attended our clinic who came from Sub-Saharan Africa. Of the 286 eligible patients 70 consented to have an HIV test of whom 16 (23 per cent) were positive. Eight of 16 were heterosexual and of those eight, seven were female. None of the women used drugs intravenously and none knowingly had had sex with bisexual men. All of the women had received injections for varying illnesses or vaccines at some point and one had had a blood transfusion. That being the case, they may well have become infected with HIV via contaminated needles or blood products. Alternatively they may have become infected at sexual intercourse.

On a global scale, however, degree of promiscuity and whether there were concomitant STDs such as syphilis or genital herpes was, and continues to be, more important than sexuality as a risk factor per se for transmission of HIV. The presence of genital ulcers promotes influx of inflammatory cells which become an easy target for HIV to infect.

Guzman, N and Vijayan, V. HIV-associated Lipodystrophy. 2021 May 12. In: StatPearls [Internet]. Treasure Island (FL): StatPearls Publishing; 2021 Jan–. PMID: 29630235.

De Vincentis, S, Tartaro, G, Rochira, V and Santi, D. HIV and Sexual Dysfunction in Men. Journal of Clinical Medicine. 2021 Mar 5;10(5):1088.

Goldmeier, D, Kocsis, A and Wasserman, M. Sexual dysfunction in women with HIV. Sexually Transmitted Infections. 2005 Aug;81(4):284.

Murphy, S, Breuer, J, Chinn, R, Goldmeier, D and Harris, JR. HIV infection among sub-Saharan African patients in the UK. The Lancet. 1989 Nov 18;2(8673):1225.

Lambert, H and Wood, K. A comparative analysis of communication about sex, health and sexual health in India and South Africa: Implications for HIV prevention. Culture, Health & Sexuality. 2005 Nov–Dec;7(6):527–41.

Wood, K and Lambert, H. Coded talk, scripted omissions: the micropolitics of AIDS talk in an affected community in South Africa. Medical Anthropology Quarterly. 2008 Sep;22(3):213–33.

Chapter 20

SEXUAL MEDICINE AND THE ORTHODOX JEWISH COMMUNITY

Tricia had embraced the life of an orthodox Jewess from the early 1980s. Her newfound belief and faith, starting from ground zero, was staggering. Judaism being very much a religion of the family, I therefore had tough decisions to make about my own level of religious practice. The older three children had been taken from a regular secular primary school and sent to an orthodox Jewish school. All of the children from then on had both a secular and an in-depth Jewish education. Some eventually went to university. One of my sons has recently completed a PhD in pharmaceutical chemistry. They are all now happily married and content in their strong religious beliefs and practice. My own parents did not talk about belief, nor did they encourage engagement in religious philosophical debate. From an early age I quite enjoyed the daily, weekly and annual cycles of practice, without giving much thought to the ideologies that underpinned the rituals. The Sabbath or Shabbat in Hebrew, starts on Friday night. During weekdays, life could be hectic, and my childhood family members dispersed. But on Shabbat we all sat together. We did not use private or public transport on that day. It was a family time, with traditional Ashkenazi food – chicken soup, fish, chicken. Passover too was exciting, when we changed over all our cutlery and crockery to a non-leaven contaminated set. We ate only Matzo – or non-leavened bread. My father led the so called Seder night on Passover, when we retold our history of being slaves in Egypt and our later freedom. My parents shared these Passover evenings in the 1950s with American GIs who were stationed in the UK, and far from home. One soldier named Gilbert, who very much appreciated our hospitality, thanked the family by taking me to a football match.

In my early teens I again began pondering what lay behind these rituals. Science told me, at that time, that the Universe was millions if not billions of years old. I found this difficult to square with the biblical age of the earth which is just shy of 6,000 years. And where was God when he allowed the Nazis to kill 6 million of my brethren? Was the Jewish God the same as the Christian or Muslim version? The lines 'How odd of God, to choose the Jews' came to my attention. To be true, it is rather odd. As a latter day wit continued, 'Not odd of God, the Jews chose God'.

Try as I might, my agnosticism was large and unmoveable. In my adolescence, a rabbi had advised me that faith comes only after ritual is practised. So aged 14 I asked my mother if she would buy an urn that gently boiled on Shabbat without kindling a light (making fire is a biblical prohibition). My mother did not take kindly to this idea. My request was refused. She would not explain why. That thoroughly disappointed me. How else could I obtain more faith? 'I know,' I excitedly informed myself. 'I will force God into my brain and mind'. I sat for many an hour, teeth gritted. I was a blinkered and troubled adolescent looking for a tenuous contact. It was rather like two major roads coalescing into a country lane. There was not space for both religious and logical belief traffic. The battle went on for two years. The logical self's thoughts raced on apace, no effort involved there. Religious aspirations were gridlocked as the scientific dialectics streamed though. My efforts were unsuccessful. And so my foray into religiosity again faded into agnosticism, but a Jewish version, where I was very much affiliated to my people and its rebirth in Israel after millennia of persecution.

So there I was, having given Tricia advice about how to live an orthodox life, with me leagues behind her. She and the children seemed very content in their religiosity. Logic dictated I could make one of two moves. I could take a firm stance and adhere to my (non) beliefs. I knew that would splinter the family and cause confusion among the children as well as marital strife. No, I was too emotionally attached to Tricia and the children. That would not do. So I took on the outward trappings of an orthodox Jew, praying three times a day, taking off all the Jewish holidays as leave, strictly adhering to the dietary laws, which meant only eating outside the home in kosher restaurants. The final icing on the cake was to grow a full beard, and wear a trilby hat on Shabbat. I even took my older two boys to see the Lubavitcher Rebbe, Rabbi MM Schneerson in Brooklyn in 1982. I respected his writings as he was a qualified engineer as well as a very holy man. Some of his opinions made sense in terms of science and religion, others did not, e.g. he insisted that taking into account the Special and General Theories of Relativity, the sun can be considered to rotate round the earth. This pre Copernican notion did not sit happily with me. I never did ask him about the physics of Joshua making the sun 'stand still in the sky'. Best left alone. All things said he was a very special man, giving mostly inspired advice to Jew and non-Jew alike. I hoped, as the earlier Rabbi had suggested and promised, that practising the religious observances would lead to faith. But the scientific mental chatter remained too strong.

As my children grew up and had families of their own, my educational role model task gradually waned. I had done my duty. But when I am with Tricia even nowadays, I am as outwardly orthodox as she wishes me to be.

And so the orthodox Jewish community in the UK, rabbis and patients, came to consult me about sexual problems, knowing that any treatment would be undertaken within what was permitted by Jewish law. Orthodox couples, at least in the past, found many non Jewish and non orthodox therapists to be outwardly hostile to what they perceived to be outdates laws and customs.

The great majority of referrals, both NHS and private were in young married couples.

Some background about sexual matters in Jewish life. There is a biblical commandment on the husband to be fruitful. He must also ensure his wife enjoys the pleasures of sexual intimacy. A wife does not have to ensure her husband's pleasure. She is a source of blessing to her husband. The Talmud, the commentaries on the Bible, state clearly and emphatically that God counts a woman's tears, and men are warned to make sure they do not cause their wife pain. Rabbi Jonah Gerondi, who lived in 13th century Catalonia, wrote that a man has no right to ever hit or abuse his wife. Jewish law deems having intercourse with one's wife against her will to be a criminal act. This notion predated that in countries in the Western world by many centuries. The UK legislated against marital rape only in the 1970s.

In the UK, orthodox Judaism is practised by a number of different subgroups. At one end of the continuum are the ultra orthodox and Chassidic Jews. Men in these sects often wear 18th or 19th century garb, particularly on Shabbat. At the other end are the modern orthodox who have assimilated secular culture into their religious framework. There has often been no sex education for either of the couple with sexual problems. They learn early on that solo masturbation is prohibited. Niddah, which is a period of physical separation from each other, starts with the onset of the menses and continues for seven days after menstrual cessation. This usually means that intercourse can be recommenced by day 12. Before intercourse the woman immerses herself in a ritual bath, termed a Mikveh, a specially constructed pool of natural water. The period of sexual abstinence may sound draconian, and indeed it is difficult for some couples. However, there are certain pluses. During the niddah time the couple can continue verbal intimacies, and by the time intercourse is recommenced the couple are desirous after the abstinence. Having sex on tap all the time can of course lead to boredom.

Unconsummated marriages, of which more later, are often underpinned by the couple's poor understanding of genital anatomy and physiology as well as overcoming pre-marital modesty.

Many referrals came to me from rabbis. I was de facto treating a threesome, with some couples reporting back to their rabbi to obtain approval for what I had suggested.

All the above stated, the vast majority of orthodox Jewish couples have good sex lives.

Benzion and Batsheva were a recently married couple who were now practising modern orthodoxy, but had received a more traditional schooling. They came to see me because they were unable to consummate their nine month marriage. It took a few sessions to disentangle the reasons for this. Benzion's erections were initially sustained until he attempted to penetrate his wife. Now when they attempted to penetrate he could no longer obtain an erection. Both had never had any sex education, had never self stimulated prior to marriage, and were unsure where the vaginal orifice was located. Batsheva had never used tampons, so that her whole area from pubis to anus was a mysterious, undiscovered land. The sequence of events when they went to bed entailed a few seconds of kissing followed by the attempted mechanics of inserting his penis into her. My first task was to teach them basic sexual anatomy and physiology, which I did in the sessions. While this was ongoing I instructed them to undertake non genital sensate focus. As detailed before, this entailed gently massaging each other in turn anywhere except the genital areas, noticing the quality of the sensations in a non-judgemental fashion, and not trying to achieve anything. This was requested of the massager and the one being massaged. I also gave them mild paradoxical instructions to, where possible, minimise erections. Of course, this had the opposite effect and took the strain off Benzion trying to obtain erections. Batsheva needed more in depth anatomy lessons. She attended on her own a few times, where I and a female nurse took her through her own anatomy, with her lying on an examination couch, undressed from abdomen down. With the use of a large mirror she was able to appreciate where the vagina was located, and where the non-engorged clitoris was to be found. After I had inserted a well lubricated little finger into her vagina, she was delighted to find that she could painlessly repeat what I had done. Next I needed her to know how to arouse herself. Female sensitive erotic literature was not appropriate, so I asked her to gently stimulate her clitoral area at home on her own, thinking about being with Benzion. This request had to be OK'ed by their rabbi who duly reported back to her that this presented no problem. Two weeks later, at the next conjoint session, Batsheva reported back that she had been able to orgasm on solo stimulation. The next stage for the couple was genital sensate focus where they took it in turns to gently massage non-genital areas, followed by ten minutes of non-goal orientated genital massage. Benzion immediately looked downcast, expressing grave concerns that this was not permitted as he might ejaculate outside the vagina. Back to the Rabbi, who told them, as the end goal of therapy was for Batsheva to conceive, it was permitted.

I told the couple to go away and practice at least three times a week. At our next meeting, they informed me, rather sheepishly, that they had broken the rules, had very enjoyable foreplay and had penetrative sex a few times. All three of us were delighted. She fell pregnant two months later.

Chani was a 40-year-old orthodox Jewish woman who lived in Switzerland. Her husband and her came from an orthodox Chassidic background. She had and continues to have a very happy marriage and is the mother of seven children all of whom were well. Unlike Benzion and Batsheva, she and her husband had been sexually inventive and had educated themselves by reading and 'learning on the job' as she put it. A year before I first saw her she was having foreplay with her husband and becoming pleasantly sexually excited. She suddenly noticed gurling noises coming from her chest and soon after became very short of breath. She was taken to her local hospital by emergency ambulance where she was found to be suffering from pulmonary oedema, that is fluid seeping into her lungs. She was treated with a diuretic which removed excess fluid from the body as well as a blood vessel dilator which lowered the blood pressure which at the time was significantly raised. The whole episode was understandably extremely frightening for her. This condition had nothing to do with her religious background. I detail her case out of interest, as well as to show that many orthodox couples, with no prior sexual experience, end up having great sex lives.

At the hospital, the doctors explained to her that because in her case the pulmonary oedema occurred suddenly, it is termed 'flash pulmonary oedema'. There are a number of precipitating events that can induce this, that include anxiety, sexual arousal and heart issues. She was later found to have sustained high blood pressure. Her Swiss physician felt strongly that she might have a benign tumour that secretes adrenaline like substances called a phaeochromocytoma, which can arise almost anywhere in the body and could be the underlying cause of the sustained high blood pressure and flash oedema. He therefore put her onto specific adrenergic blockers (alpha and beta). She had repeated biochemical tests and MRI scans to look for a phaeochromocytoma. However to date no tumours have been located.

Chani and her husband were, and still are understandably very apprehensive about having sex. They miss each other's bodies and mutual physical affection.

Each time they just went further than a fond peck kiss, she heard the gurgling and then just stopped. Her blood pressure is still very variable, but can at times be very high and seemingly unresponsive to medication. She had a repeat mild and transient episode of flash pulmonary oedema when she was involved in a minor car accident. In sessions with us she engaged in mindfulness and practised it diligently. In spite of all therapies, she is still unable to safely have sex with her husband. As one of the physicians looking after her remarked, 'We must be missing a trick with her'. That said, there are a number of cases in the literature of recurrent severe flash pulmonary oedema where no gross abnormalities are found apart from the episodic excess fluid in the lung. In many cases high levels of anxiety and sexual arousal are presumed to precipitate the attacks.

Pesach was a 40-year-old man who came from a Chassidic background and had been raised in New Jersey. He and his wife had been introduced to each other by a matchmaker (shadchan in Yiddish) and found each other reasonably compatible after two short meetings, and so went ahead and married. Pesach's first language was Yiddish. However he spoke passable English. He looked so sad as he sat, long, slightly gingery, curled sidelocks appearing as he took off his hat. The biblical commandment of growing sidelocks, or peyot in Hebrew, is connected to not worshipping idols. He and his family had been told to move from New Jersey to London because of his sexual problem. In spite of having a wife and children, he felt he lacked something in his marital life. That being so, he began texting, on an almost daily basis, an 18-year-old unmarried woman who he had first spoken to quite casually in a shop. He had met her face to face every month or two, and then only to walk with her in a wooded area. The parents of the girl were astonished that this married, seemingly upstanding stalwart of their community, had engaged their daughter in such a shameful way. He agreed to stop so doing, but after three months recommenced. His rabbi insisted he see a modern orthodox endocrinologist who put him onto antiandrogens, that is medication to cut his sex drive. He nevertheless continued to intermittently text the young woman and see her sometimes. Fortunately the medical treatment did not seem to prevent him having intercourse with his wife, which he said was satisfactory. His rabbi thought the best move was to relocate his family to London, from where he continued the texting but of course could no longer meet her. I asked him if his wife would attend our sessions. She refused. I pressed on, not being sure what I was dealing with here. I asked for more detail of what happened in the marital bed. How was foreplay? He did not understand what I meant. It seemed that when they had sex, there was no foreplay at all, no kissing, just straight to penetration. That is what he was told to do before marriage. How did they communicate with each other? Was there tenderness? It transpired that his wife, for reasons I never found out, could not express any loving or tender feelings towards him. So, Pesach did not have a sexual problem. What he lacked was warmth and intimacy, which he found in the young woman. He went on to discuss this lack with his wife. He left therapy before I could discover the outcome.

Mendel lived in Monsey in upstate New York. He came to visit a rabbi local to me, but had a problem that related to cholent. Cholent is an Ashkenazi Jewish food that is much like a stew. It is eaten on Shabbat lunch. Because of the restrictions on active cooking on the Sabbath, it is prepared before the start of that day and put in a slow cooking crock pot to simmer and gently cook overnight. In past centuries it was put in a baker's oven that remained hot. Alternatively it was put in a haybox to cook. Cholent usually contains a grain, vegetables and beef, with a liberal sprinkling of spices. By the time it is eaten the food has an pleasurably intense odour and flavour. The etymology of the word is obscure, but might derive from the old French word Chalant meaning to warm, or a combination of later French words chaud (hot) and lent (slow). Mendel's wife was at her wits' end. She and her husband had been married for ten years, and had no children. No surprise in that she claimed, as he was not interested in having intercourse with her. But on Shabbat lunch when the cholent pot was placed in front of him, his eyes glowed, and to his wife it appeared he was attracted to either the cholent or the pot it was contained in. Her anger on one of these occasions overflowed. She put her hand on his groin and found he was erect. 'You find the cholent beautiful but not me,' she screamed at him. He was silent and ashamed, and did not reply to her outburst.

Mendel and the rabbi wanted me to wave my magic wand. I asked him if he found his wife attractive to which he answered in the negative. But he certainly had a strong fetish for the cholent. I advised them to see a particular psychiatrist in New York City with a view to discussing his marriage and his fetish.

Friedman, S. Assessing and treating sexual dysfunctions on orthodox Jewish couples: a summary of 41 consecutive cases. Mental Health Religion and Culture. 2019;22:930–42.

Wilmshurst, PT. Pulmonary oedema induced by emotional stress, by sexual intercourse, and by exertion in a cold environment in people without evidence of heart disease. Heart. 2004 Jul;90(7):806–7.

Chapter 21

NA MEETINGS AND SEXUAL ADDICTION

I continued to attend NA meetings for almost 25 years. Some would say that was long enough. The 12 step NA hardliners, however, would advocate engagement for life. Addicts never complete the process of recovery. They are perpetually 'recovering'. So introducing myself at meetings I would say, 'My name is David and I am a recovering addict'. The folk I met in 'the rooms' were never dull, and I learnt something from everyone I came across. The attendees ranged from addicts who had over 30 years' clean time, to others who came into the meetings stoned. A psychologist who himself was in recovery undertook research that showed it took at least three years of clean time to attain any semblance of psychological, spiritual and relationship normality. There were many who could only attain a few months, weeks or even days clean. Some died of overdoses, which upset us, but was a salutary warning of what could happen to any of us if we returned to our old ways.

Certain recovery phrases were heard time and again at meetings:

'Recovery is a day at a time'

'One drug is too many and a thousand never enough'

'Hugs not drugs'

'This too shall pass'

'Live life on life's terms'

NA meetings were a social strata leveller, so that a rich professional newcomer who was willing to change could learn from an unemployed member who had many years' clean time and had been through the steps multiple times. A vital part of the recovery process was 'doing service'. There were a number of such positions at any meeting. If there was a vacancy one could put oneself forward for the role. If more than one person volunteered, a secret vote was taken. The 'tea person' was a vital job. It entailed making sure that tea, coffee, milk, biscuits and sugar were available to addicts as they entered the building before the meeting. The tea person was in effect a greeter, who particularly kept their eyes open for the shy newcomer. An older man with many years' clean time often recounted how when he first got clean he lived in New York. His family had disowned him. He had no money but survived for a few weeks by eating biscuits and drinking coffee in a warm room, going from one meeting to another in the city for 24 hours each day. Funding for the tea person's victuals, paying the room's rent, buying literature and forwarding money to NA UK or World NA was by donation either at the start or the end of each meeting where 'the pot' was circulated. The person who requested money from the meeting was the Treasurer, whose responsibility it was to pay the rent for the room – often in a church hall. The meeting's Secretary read the introductory and wrap-up formats. If a meeting became verbally aggressive, disputatious or personal she/he could intervene, even closing the meeting early if that was what they considered best for recovery. After some readings from NA literature a 'chair' person would tell their story, usually incorporating how working through the Steps had aided recovery. Occasionally a relative newcomer would merely list their downward journey to rock bottom, listing what drugs they had taken and how much they used, without much of a recovery story. Such a tale was deemed a 'drugalogue'. The chair took 10–15 minutes after which other members would reply, sharing their experiences of recovery. Another service position was the GSR or group service representative, who would go to a meeting in the larger area where matters of policy would be discussed. When I was GSR I found the area meeting chaotic with much bickering. Over the course of the years I undertook all of these service positions. Meetings have changed since the end of the 1990s, when sessions would invariably be smoked filled. Nowadays NA literature and details of all UK meeting venues and times are available online. In the past these details were solely found in little white 'Where to Find' booklets. Meetings in a church hall in Willesden could be interrupted by the needy and neglected from the nearby soup kitchen wandering, as well as by noisy film crew shooting horror movies in the nearby churchyard.

So, did I have an urge to restart codeine? Of course the thought crossed my mind. I recalled how good I felt when first taking it. But I would lose all my recovery time which I was becoming more and more proud of as the years rolled by. At the anniversary of one's drug cessation one was given a 'Clean time' key ring plus lots of hugs and claps. And more to the point, could I go through opiate withdrawal again? Most definitely the answer to that was no, although prescribed opiates awaited me in the not too distant future.

Occasionally doctor addicts attended NA meetings. A young anaesthetist asked if I would sponsor him and take him through the NA 12 steps. He had not come willingly into NA, but had his General Medical Council license to practice withdrawn. He had a real issue with Step 2, which states, 'We came to believe that a power greater than ourselves could restore us to sanity'. Gilbert undertook private work doing terminations of pregnancy under light anaesthesia with the aid of the powerful opiate fentanyl. Before the session he would draw up ten syringes of fentanyl if there were ten procedures to carry out. However if one or more patient did not show, he deemed it a waste not to use the contents of the remaining syringe. So rather than bin the unused drug, he injected their contents into himself. In spite of being a highly intelligent man, he could not perceive the insanity of his actions. He only attended NA and only had me sponsor him because he was told it would have to be part of his recovery programme. There was a lot about him that I never got to understand, nor did he for that matter. He was not in a regular relationship but rather had casual sex, often with many partners over weekends or holidays, when he became bored. Rather than address the cause of his boredom he either masturbated to porn or had condomless sex with female casual partners. He told me he found it difficult to find time to write out the steps, but in reality he was merely treading water and not engaging with the NA programme. I lost touch with him after he moved to Wales and later on to the USA. I do hope he managed to find some sort of insight.

In 2011 Jenny Petrak, a clinical psychologist and I wrote a short paper on sexual addiction in a journal specialising in STDs. We discussed the fact that in that clinical setting there are obvious adverse consequences of being with multiple partners, including STD acquisition, unwanted pregnancies and relationship distress. We quoted Martin Kafka, a recently retired psychiatrist from Harvard University, who had reviewed the various models and nomenclature of so-called sexual addiction. He uses what I believe is the helpful term 'hypersexual disorder' (HD), which occurs in non-paraphilic patients, where the term paraphilic defines a sexual interest in an atypical target e.g. articles of clothing, children. Kafka's diagnostic criteria for HD include uncontrollable sexual urges or actions (e.g. watching porn) which interfere with other aspects of a person's life such as work or personal relationships, engaging in these actions because of a mood problem e.g. anxiety, depression or boredom, which are often due to stressful life events, trying unsuccessfully to stop the repeated actions, and disregarding any adverse outcomes of the activity. There needs to be distress felt by the person. However misuse of alcohol or drugs excludes a diagnosis of HD. Having multiple partners per se does not imply HD, nor does watching porn for say 30 minutes a day if the other criteria above are not satisfied. Treatment options can include mindfulness based CBT, group and family therapy, attending 12 step programmes such as Sex Addicts Anonymous and judicious use of antidepressants such as SSRIs. Kafka's classification was not accepted as a diagnostic category in the American Psychiatric Associations recent 5th version of the *Diagnostics and Statistics Manual* (DSM 5). Whatever the diagnostic criteria used, these patients are a problem to themselves and the community, such that half the reported STDs reported in the UK over a recent five-year period occurred in less than the 10 per cent of the population who had sex with ten or more partners.

Douglas Braun-Harvey, a psychotherapist working in San Diego, takes a rather different stance on what appears to be sexual addiction, particularly in gay men. He encourages careful differentiation between compulsive or problematic sexual behaviour in gay men on the one hand and developmental, situational or culturally normative sexual behaviours on the other. He feels that therapists would be more effective by conveying interest and curiosity about their clients' sexual activity and other presenting concerns. He advocates individual, couples and group psychotherapy for what he describes not as addiction but as 'out of control sexual behaviour'.

David Marcus, a psychologist in Silicon Valley California prefers the term problematic sexual behaviour. Rather like Kafka, he differentiates between libido driven sexual urges and anxiety driven sexual behaviour. The initial trigger, is often avoidance of anxiety. He, like the other experts, takes a multimodal treatment viewpoint but advocates individual psychodynamic therapy as the basic vehicle for change.

Caleb was a 30-year-old gay man who worked in publishing. His five-year relationship had become non sexual three years before. Caleb's anxiety had been problematic since childhood during which he suffered from difficulty in getting off to sleep and bad nightmares. He did reasonably well at school and went on to study English at a good university, ending up with a 2.1 degree. After this he worked at a well-known book publishing house, starting out as a proofreader. He subsequently progressed to a more senior supervising role, but found this really stressful, complaining of palpitations, shaky hands and severe restlessness. Getting off to sleep had been a problem for the previous two years. At about the same time he found that solo masturbation, when he 'edged' near to orgasm while watching porn, had given him some relief from the anxiety and worries. He spent 3–4 hours a day at home masturbating. Because of this he had many absences from work that had come to the attention of his superiors. He had tried many times to cease watching porn, but his anxiety had immediately resurfaced.

When seen at clinic he denied using drugs or alcohol. He was obviously very distressed about his problem.

The concept of hypersexual disorder was discussed with him. Treatment was with an SSRI antidepressant fluoxetine at high dose, as well as mindfulness based cognitive therapy. Four weeks later his anxiety had improved considerably, and he was back at work full time. Mindfulness helped in that it reduced anxiety but also encouraged sitting in uncomfortable moments, being an observer rather than at the mercy of the discomfort.

Kafka, MP. Non Paraphilic Hypersexuality Disorders in Principles and Practice of Sex Therapy. The Guilford Press, 2014. Eds Binik, YM and Hall, KSK. pp 280–304.

Braun-Harvey, D. Culturally relevant assessment and treatment for gay men's online sexual activity. Sex and Relationship Therapy. 2003:18:371–84.

Goldmeier, D and Petrak, J. How to recognise sexual addiction in the sexual health clinic setting? Sexually Transmitted Infections. 2011 Aug;87(5):370–1.

Fenton, KA, Korovessis, C and Johnson, AM et al. Sexual behaviour in Britain: reported sexually transmitted infections and prevalent genital Chlamydia trachomatis infection. The Lancet. 2001 Dec 1;358(9296):1851–4.

Marcus, ID. Problematic sexual behavior. In Handbook of Clinical Sexuality for Mental Health Professionals. Routledge, New York Third Edition, 2016. Editor Stephen Levine pp 164–77.

Chapter 22

THE JANE WADSWORTH CLINIC, ITS MEETINGS, PATIENTS AND LECTURES

Into the new millennium I was spending over half my week working in sexual dysfunction alongside my two psychologist colleagues Agnes Kocsis and John Green. Willie Harris, my physician colleague, was also doing some sessions in sexual dysfunction. A senior nurse spent a session training men with ED to give intrapenile injections, as well as following up patients who the physicians had initiated on oral medications. I obtained NHS funding for a pelvic floor physiotherapist. The first of a number of women who held that post was Julia Muman, whom I sent to Montreal to work with Irv Binik's 'physical therapist' (as the speciality is named in North America), to learn how pelvic floor work could help patients with sexual problems. I had, by this time, succeeded in carving out a stable niche for the treatment of sexual dysfunction management within the greater sexual health setting of the Jefferiss wing. My thoughts roamed. Sexual medicine at St Mary's was now well known at least at a national level. I pondered how I could further enhance its reputation. A part of me, truth be known, continued to feel ridiculed and marginalised by the bygone reputation of Venereology. What I was trying to achieve was to reshape and remodel that bedraggled and soiled status of Venereology, transforming it into respectable GU medicine, and onwards into sexual medicine in particular, a speciality that I hoped any hospital would be proud to be associated with. I therefore decided to give the clinic an eponymous title. Medicine is littered with diseases with strange eponymous titles. Munchausen's syndrome is named after the German aristocrat who exaggerated his adventures. His syndrome refers to patients who give untruthful and dramatic accounts of their fictitious illness. I have already discussed Lou Gehrig's disease. Fictitious characters may have a syndrome named after them, such as in the Peter Pan syndrome, where the adult, for a number of reasons, remains dependant and does not wish to be an adult. But after whom should I name the sexual dysfunction service? After a few days of roaming thoughts and musings, I decided it would be suitable to name it the Jane Wadsworth Clinic. Jane Wadsworth had been a medical statistician. She had studied mathematics at St Andrews University in Scotland and then began work in London's Institute for Social Sciences in Medical Care. She later completed a course in medical statistics at the London School of Hygiene, followed by accepting a lectureship in medical statistics at St Mary's Hospital Medical School in 1983. She became involved in HIV epidemiology at that time, which laid the foundations for sex research which she carried out with Anne Johnson and Kaye Wellings resulting in the first National Sexual Attitudes and Lifestyle (NATSAL) survey which was published in 1994 (there have been two subsequent NATSAL surveys at ten-year intervals). Although clinicians can find talking to medical statisticians daunting (me in particular), this was not the case with Jane who always made the problem at hand easy(ish) to understand. She was also very approachable. She died of a rare lung tumour in her 50s. I approached her family who were delighted and honoured that Jane would be remembered.

In 2001 I approached St Mary's Hospital NHS Trust Chief Executive and Chairperson to endorse this clinic naming. They too were enthusiastic. Together we organised an afternoon event, consisting of lectures and culminating in a tea reception. Jane Wadsworth's family were present. The event took place in the Cockburn lecture theatre, named after Dr Henry Cockburn, a brave World War 2 Army doctor and later Medical Superintendent of St Mary's.

I invited a number of dignitaries from the UK sexual medicine field. The lecturers included Alan Riley who held the chair of sexual medicine at the University of East Lancashire. I can't recall his topic exactly, but it may have included experiments to determine where female lubrication at arousal derives from by using cotton wool pledgets placed at strategic points in the lower genital tract and weighing them before and after sexual arousal. By so doing he found that about a third came from the Bartholin's glands opening into the vulva with the rest a vaginal transudate.

Chris Mathias, who held the chair in neurovascular medicine spoke about sexual problems in patients with autonomic neuropathy (disease of the nerves that automatically control the function of the heart, lungs and gut). Dr Onn Min Kon, as he was then (later became Professor) spoke about sleep apnoea (periods when breathing does not take place during sleep) and sexual problems and Agnes Kocsis about sexual therapy. There was a talk about Jane Wadsworth given by Anne Johnson. After the formal part of the day there was a 90 minute reception. The hospital Trust board paid for the afternoon's expenses – not sure this would happen in 2024!

From 2001 onwards the Jane Wadsworth team met for an hour or two on most Wednesday mornings, to discuss patients at a combined multidisciplinary team meeting. A few times a year we invited speakers to lecture on topics of interest. I have documented below some cases and themes of interest that arose in meetings in the following paragraphs.

Fertility, clitoral and vaginal orgasms

Dr Roy Levin, who until his retirement was a Reader in Physiology at the University of Sheffield, gave us a talk on clitoral and vaginal orgasm. He was an excellent speaker, who also had a passion for collecting Roman ointment jars. He told me that, not infrequently, he found them going for a song at markets and antique fairs. He had written extensively on the topics of clitoral and vaginal orgasming for nearly 40 years. In the new millennium he authored a number of highly precise and detailed articles in scientific journals heavily criticising the work of Stuart Brody (see below).

Rose was a 30-year-old fitness instructor who had been in a relationship with Cyril for a year. We had initially seen Cyril because of psychogenic ED which had now resolved. The couple were not using contraception as both wanted to have children. However after a year she had not become pregnant and was waiting for an appointment in a local fertility clinic. Rose had attended our department earlier on as Cyril was concerned not only about his ED but also worrying that Rose orgasmed clitorally but not vaginally. He felt if she could have a vaginal orgasm she would enjoy sex more and would also have a higher chance of conceiving. Rose felt very criticised by this, and reluctantly had taken on trying to achieve a vaginal orgasm, feeling her lacking in this department was the direct cause of her not conceiving. She looked at sexually arousing videos and inserted penile shaped vibrators to encourage a vaginal orgasm but to no avail. We assured her that it was quite normal for some women to be able to orgasm vaginally and some not. Additionally they both wondered if her not standing up after his ejaculation, so as to not let semen trickle out of her, would be helpful. We told the couple that there was no evidence that remaining supine after ejaculation was helpful in terms of conception.

Three months after first coming to see us Rose conceived by natural means.

Masters and Johnson in the 1960s took video film of women in sexually aroused states and found that at orgasm, the uterus dips into the vagina with each rhythmical contraction. It was suggested on this basis that semen previously deposited could be 'upsucked' into the uterus and so facilitate conception. This however is not the case, as sperm that have just been released at ejaculation are incapable of fertilisation. After deposition of semen in the vagina, the sperm become coagulated for up to 15 minutes, after which there are biochemical changes within particular sperm, enabling them to be reprogrammed or 'capacitated'. It is these capacitated sperm that are able to penetrate the ovum and produce fertilisation. Therefore whether or not a vaginal orgasm takes place 15 minutes prior to capacitation is irrelevant to conception. Likewise semen that leaks out of the vagina on standing up after sex has not been coagulated so cannot lead to conception. It is well known that women neither need to be aroused or to orgasm in order to conceive.

In 1905 Sigmund Freud put forward the notion that a woman's orgasm achieved by clitoral stimulation was a reflection of an immature psyche and that true femininity only occurred when a woman was able to achieve orgasm purely from penovaginal stimulation. This hypothesis has been heavily censured over the years and indeed there is little empirical evidence for it. However, recently, Stuart Brody, a psychologist now working in Prague, has resurrected this theory with a number of publications suggesting that women's orgasm purely produced by penile thrusting has, in addition to being associated with 'psychological maturity', other beneficial effects such as more positive attitudes in general and better general health. His work has also been heavily criticised, particularly by Roy Levin, who points out that Brody's studies purely show correlation rather than causation and have not been repeated as yet by others. Levin points out that arousal in women is rarely restricted to the clitoris or vagina alone, but often both are involved in a partner's sexual stimulatory activity. A number of studies have shown that the anterior or front wall of the vagina, wherein lies the so-called G spot, as discussed above, in reality also houses the body of the clitoris, so that when vaginal thrusting takes place the clitoris is being stimulated. Furthermore ultrasound studies show that at penovaginal thrusting, a greater part of the glans of the clitoris near the urethra is dragged in and out of the vagina and so is being stimulated, implying that there is no such anatomical event as pure penovaginal sex.

Levin, RJ. The human female orgasm: a critical evaluation of its proposed reproductive function. Sex Relationship Therapy. 2011:26:301–14.

Sexual surrogacy

Sexual surrogacy was a topic hotly debated at our MDT meetings.

Sexual therapy with couples requires cooperative partners who understand the therapeutic process. Surrogate partner therapy (SPT) provides a solo individual in sex therapy with a proxy partner who can take him or her through the therapeutic process. It also enables patients to engage in a meaningful relationship, albeit a transient one. Surrogates assist the sex therapist, and act as mentors, with the patient developing social and intimate connections, reducing anxiety and increased confidence. Although practised legally in a few countries (e.g. Israel) it still has a poor reputation in many circles. Masters and Johnson mentioned sexual surrogates in their writings in the 1960s. Martin Cole, who was a sex education advocate and abortion law reformer started a surrogacy service in Birmingham in the 1970s. A local MP at the time described his work as akin to keeping a brothel. In Israel a surrogacy service currently trains and provides sexual surrogates for use in sex therapy clinics. In Israel surrogates are screened carefully and then have a 40-hour training endorsed by the Professional Surrogates Association. They do not need to be particularly attractive. Initial sessions with the patient and surrogate may involve just getting to know each other. Surrogate therapy is really a tripartite alliance between the therapist, surrogate and the patient. The conclusion of the therapy signals the end of the surrogate patient relationship.

In the UK currently there are at least two surrogate-like therapies available. One is at ICASA (Intimacy, consciousness and self-awareness) where the therapist is David Brown, who is a 'Master of Metaphysical Science', and a member of the Interfaith Ministers Association, and a therapist member of the International Professional Surrogates Association. To my knowledge he does not encourage gay patients attending his unit, which is sited on a farm in Bedfordshire.

Mike Lousada a psychologist practises 'Bodywork' in North London. Sessions with him are generally three-hours long. Genital hands-on work is only done with him if deemed appropriate and is discussed in advance with the patient. With him the therapist and 'surrogate' are melded into one.

Ansh was a 38-year-old actuary who had suffered from premature ejaculation (PE) since acquiring psoriasis some five years before he originally saw me. Prior to that time his intravaginal ejaculatory latency time was up to ten minutes. His latency time was now less than a minute. The likely cause of his secondary PE was prostatitis, possibly associated with the psoriasis. I confirmed the prostatitis by ultrasound and tests of his prostatic fluid and urine. After receiving a long course of antibiotics he was no better. SSRI antidepressants made him feel ill. Local anaesthetic creams were helpful to the extent he could continue penile stimulation for two minutes at solo sex. He had only had one long-term relationship with Sara some ten years before. With casual partners he took Viagra which gave him confidence in terms of erection hardness and enabled a second erection soon after the first orgasm at which he could last for two minutes.

He decided that what he needed was a sexual surrogate and to that end made a lot of enquiries in and around the red light district of Amsterdam, finally meeting a young woman who said she could help him. He found her attractive.

Each session with her lasted 90 minutes. He did not feel rushed. Using the 'stop-start' technique and proceeding to vaginal intercourse he was able to continue vaginal sex for five minutes albeit with a condom. He told me that he really would have liked a relationship with her but when he asked her she replied that this would be stepping beyond her boundaries as a surrogate.

Nawaz was a 50-year-old Pakistani man who had been in the UK for 20 years. He held a responsible job at a major supermarket but lived on his own. Much of what he earned he sent back to his family who lived in a rural area of Pakistan. When he was eight he was innocently playing with girls and boys of own age. This involved the small group briefly looking at each other's genitals. Midway in one of these sessions an adult found them and angrily blamed Nawaz for what they were doing. The adult told Nawaz's parents as well as the rest of the village. His shame knew no bounds. This event played heavily on his mind throughout his life. As an adult in the UK he could go out in social groups, but was too shy to ask any woman for date, in part because he had developed erectile dysfunction as a result of diabetes some ten years before, and partly because of the traumatic memory. Before he came to see me he had undertaken trauma therapy. However it did not seem to help his ability to talk to women one to one, let alone ask one on a date. We taught him to use intra penile injections which gave him good erections in the clinic setting and on his own.

At this point he saw my sex therapy colleague who felt a sexual surrogate might be useful. The MDT did not feel it was ethical for us to find him a surrogate but felt it was ethical if she told him how to use the web in order to find such a woman. He found one called Mandy who was very understanding. She spent a number of sessions just getting to know him. Later sessions involved them taking their clothes off and gently taking it turns to massage each other as per sensate focus. He enjoyed these sessions. Although he injected himself prior to the session (with Mandy's knowledge) his erection often faded, likely because of his high anxiety levels. When I last saw him he had not been back to see Mandy as he was short of money, having sent a great deal to his parents.

Rosenbaum, T, Aloni, R and Heruti, R. Surrogate partner therapy: ethical considerations in sexual medicine. Journal of Sexual Medicine. 2014;11:231–9.

A shock in the attic

Agnes, my psychologist colleague and I worked a whole day each week together. A psychologically orientated doctor and a psychologist with a medical bias was a powerful therapeutic combination. We felt as a team we had the measure of most of our patients. However we were totally nonplussed by the outcome of the treatment of Bernard and Rosamund. A salutary lesson that the truth can be difficult to access.

Bernard came to our clinic with Rosamund his wife on every occasion he came to see us. He was 68-years-old, rather overweight and spoke slowly with a confident air implying that he had everything in his life under control apart from his diabetes and consequent erectile dysfunction. Rosamund on the other hand was shy, said very little but smiled nervously from time to time seemingly as an affirmation of his story. He hardly looked at his wife as he gave his account but when he did he would refer to his wife and himself as a couple. So he would say 'our penis is not working', to which she would nod in agreement. He was taught how to give intrapenile injections which resulted in quite reasonable erections. On follow up six months later he reported that 'we are happy with the outcome'.

Two years later Rosamund contacted us in an extremely distressed state, informing us that Bernard had suffered a major stroke. Between the phone call and seeing her in person a week later he had died. Before his funeral she went through the house gathering his effects. That included going into the large attic to which she had been denied access in his lifetime. He had told her that there were photographic equipment and photographs stored there and he did not want them to be disturbed. This made sense to her, as he travelled from the suburbs to central London once a week to attend his photographic club meetings. However what she found was something quite different. There were various articles of leather clothing, whips, neck collars adorned with large spikes, silver and gold looking handcuffs as well as many photographs of people including himself in various states of bondage or seemingly inflicting pain on others.

Putting all the information together Rosamund deduced that Bernard had spent one whole day a week indulging in Sadism and Masochism (S and M), rather than the more banal photographic club he had asserted he attended weekly. She further told us that Bernard had not wanted sex with her for over 20 years, and that although he was happy with the injection treatment for his ED, he never used it to make love to her. Indeed they had had no physical contact of a tender nature for all that time either. We saw her a few more times. She not only had to grieve his passing but also the loss of the husband she thought she knew but in reality did not know.

In retrospect Bernard was likely a psychological bully, controlling many aspects of his wife's life. She denied physical violence had taken place, but we wondered if she was eulogising his memory.

S and M are ways of using pain as a sexual stimulant. All activity within S and M acts involves a power exchange or power play. The submissive person controls the level of play.

A more accurate and inclusive term for S and M is BDSM (bondage, discipline, dominance, sadism and masochism). An Australian population survey found that only 2 per cent of respondents actively undertook BDSM, although many more than this have such fantasies. The mental health of BDSM participants is similar to that of the general population. One way of seeing BDSM is that it is a widening of sexual interests rather than a fixation of one generally accepted mode of sexual expression. All this stated, inflicting pain on another person within a sexual scenario that is non consensual should be seen as both a paraphilia and a criminal act.

Brown, A, Barker, ED and Rahman, Q. A Systematic Scoping Review of the Prevalence, Etiological, Psychological, and Interpersonal Factors Associated with BDSM. Journal of Sex Research. 2020 Jul;57(6):781–811.

Asexuality and testosterone

We did not see many patients who we considered to be 'asexual'. But when a candidate patient was presented there was much discussion.

The concept of asexuality is complex and still controversial. Lack of sexual attraction to others but not lack of sexual drive per se is one description. It is a solitary desire or drive. This 'lack of sexual attraction' is the generally accepted definition by the Asexuality Visibility and Education Network (AVEN). To make matters more complex, some asexual people can experience attraction in isolated instances – this is described as 'Grey A', and seems to fall somewhere between being asexual and being sexual. In the past Kinsey (in the 1950s) referred to asexuality as 'Condition X'. Lori Brotto, a professor of psychology in Vancouver, has recently considered the evidence that asexuality could be seen as a psychiatric disturbance or a sexual dysfunction and found little evidence for either. She did find that there was some preliminary evidence that asexuality might be a paraphilia, but finally concluded that there is modest support for asexuality's placement as a unique sexual orientation.

Pamela was a, tall, 39-year-old veterinary nurse with a serious demeanour. She had many friends both male and female. However she had never had a sexual partner and indeed had never had an urge to have sex with either a man or woman, and had never had sexual fantasies about anyone. She was able to describe to me that some people she knew were attractive in an aesthetic but not a sexual way. There was nothing in her personal or family history to account for how she felt. She denied non consensual sex, and had no evidence of past or current psychiatric illness. In spite of all of this, she had a drive to clitorally stimulate herself, during which she had no fantasies but enjoyed the arousal and was able to pleasurably orgasm. Her current problem was that her sexual drive had significantly diminished over the past year. She had developed hyperthyroidism two years previously, which had been managed with medication. Her total testosterone was in the low range and her sex hormone binding globulin (SHBG) very high, likely because of the hyperthyroidism. This meant the level of free testosterone, that is the amount that is biologically most active in the brain and body, was lowered. After careful and repeated testing and discussion about the evidence for testosterone's efficacy, and known short to medium term safety, she was given an appropriate dose of testosterone in transdermal gel form. After about four weeks her sexual drive returned. It seems then that Pamela was asexual. We discussed this concept with her, and she was content with it. However her sexual drive had drastically diminished. In the premenopausal state testosterone is produced by the ovaries and adrenal glands. Between a woman's mid 30s and early 60s adrenal testosterone reduces by about two thirds.

There are no current treatment licenses for the use of testosterone in women. Nevertheless giving it where needed may be very useful as illustrated by Pamela. Patients should be informed of the possible side effects of hirsuteness (hairiness) and voice changes. In Pamela's case she did develop some facial hairiness but managed it herself. Blood checks to ensure that no increase is the amount of haemoglobin (haematocrit) and blood lipids should be undertaken regularly. Major studies show no convincing evidence of an increase rate of cardiovascular disease or breast cancer in women on testosterone, but nevertheless these should be checked regularly.

Unlike total testosterone blood tests in men, which give robust results, the assay in women is fraught with problems, including low sensitivity, as the levels in women are about a tenth of men's, and vary with the time of the month. However doing the assay in the first half of the woman's cycle in the early morning and fasting plus an SHBG will give some indication of the free testosterone level and the need for replacement, particularly if the tests are repeated over time.

Brotto, LA and Yule, M. Asexuality: Sexual Orientation, Paraphilia, Sexual Dysfunction, or None of the Above? Archives of Sexual Behavior. 2017 Apr;46(3):619–27.

Parish, SJ, Simon, JA, Davis, SR et al. International Society for the Study of Women's Sexual Health Clinical Practice Guideline for the Use of Systemic Testosterone for Hypoactive Sexual Desire Disorder in Women. Climacteric. 2021 Apr 1:1–18.

Asexuality Visibility and Education Network. https://www.asexuality.org/

The cost of treating women with sexual dysfunction

As the years of the new millennium passed by, funding of the Jane Wadsworth Clinic came into question. Before that, its costs were absorbed by the general funding of GU medicine clinics. We therefore undertook a study, along with two health economists, into the cost implications of treating women.

We looked at all the women who attended our sexual function clinic over a three month period and costed their treatment. Direct medical costs were calculated which included practitioner time, laboratory investigations and any pharmacological treatments given. Of the 47 referrals 38 undertook treatment, the remainder not proceeding to follow up. The average cost per woman was £472 compared to £335 for a year for men with erectile dysfunction. The cost of PDE5 inhibitors such as sildenafil has fallen steeply since this paper was written, as these drugs are now prescribed generically rather than by trade name, so that the cost for men is now likely to be significantly lower in comparison to women. The women were seen either by a physician alone, physician with a psychologist, psychologist alone or sex therapist alone. The actual number of therapy sessions per patient varied from one session (two patients) to 51 sessions (one patient). Based on these figures, some ten years later we obtained specific monies from the Primary Care Trust to fund the Jane Wadsworth Clinic.

One of the patients who attended once only was a 33-year-old woman who worked as cabin crew on a well-known airline. Her boyfriend of two years worked as ground staff for the same airline. Sex had not been a problem since her debut aged 17, so that she had good levels of sexual desire and arousal and could orgasm clitorally and vaginally. She came to see me because she developed a moderately severe aching pain in her lower abdomen over the past six months that coincided with sexual arousal. This resolved immediately after she orgasmed but took up to three hours to dissipate if she was unable to orgasm. I noted she had varicose veins in one leg. She suggested that this was not uncommon among aircrew on long haul flights, who could be standing up for eight hours or more at a time. I could find no abnormalities on abdominal examination, but suggested she might have engorged pelvic veins causing the pain problem. Enlarged veins in and around the vagina, uterus and fallopian tubes are not uncommon. However if there is pooling and backflow in these veins i.e. they are incompetent, then they may be causing problems, in particular pain. I discussed the clinical picture with one of my radiologist colleagues, who undertook a Doppler ultrasound (US) to assess her. A bit about Doppler US – miss this out if you don't like the physics!

Doppler ultrasound investigations have been used since 1970. The Doppler effect in terms of sound relates to the sound frequency of moving objects, so that for instance a police siren has a higher pitch as the vehicle approaches, when sound waves are compressed, and a lower pitch as the vehicle recedes when the wavelength elongates. In the ultrasound investigation, frequencies higher than audible sound (20 to 200 MHZ) are projected onto a vessel. The change in reflected frequency is determined by the flow in the vessel.

We found large incompetent veins around the uterus extending to the ovaries on both sides. The treatment would be to embolise them, that is to block them, for instance with metal coils so they would no longer engorge at sexual arousal, if this indeed was what was causing her pain. In the event the patient did not reattend. I rang her. She told me that she longer had the problem. I presume the veins are still present, but seemingly no longer cause pain at sexual arousal. Strange but there it was!

Embolisation of symptomatic and incompetent pelvic veins has been reported to generally lead to clinical improvement of abdominal pain. However the presence of incompetent pelvic varicosities does not necessarily mean they are the cause of a patient's complaints.

The patient who had attended 51 times in the study was a 68-year-old physician, originally from the USA. She had first attended our clinic a number of years before, because she had never been able to have penetrative sex. She had divorced her husband of 25 years some years before coming to our department, seemingly because they were incompatible. Over the years he had psychologically abused her. As a child she recalled having to dress up in pretty clothes, only for a friend of the family to inappropriately touch her when the parents weren't looking. She had been depressed quite severely many times over the years. Antidepressants worked well. However she stopped them once the immediate crisis had resolved. From the genital point of view she was suffering from vaginismus with post-menopausal vaginal change. She was given oestrogen cream to make the vagina more supple, and started using vaginal dilators, which in her case were used to retrain her pelvic floor muscles as well as actually stretch her post-menopausal vagina. After some months she was able to insert the largest trainer. We also did some work with her to help her grieve for the pregnancies and children that she never had. Later on she felt able to self stimulate and derived some pleasure from so doing. My psychologist colleague and I encouraged her to consider embarking on a sexual relationship. She enjoyed painting classes and had met a man there who was obviously interested in her. Because he was married she did not want to pursue an intimate relationship with him. In the latter years of her seeing us, her ex husband became terminally ill. She chose to be with him and visited him often as he became weaker. In some ways this helped to resolve some of the very negative feelings that had built up between them over the years. In his will he left her his business plus his bank savings, so that in her retirement she was financially secure. Over the years her nieces and nephews had been both a joy to her and at the same time a reminder of the children she might have had herself. She continued to take her antidepressants regularly, as well as having regular psychodynamic psychotherapy which she continues to find useful. She has not made a sexual relationship over the years we have seen her.

Goldmeier, D, Malik, F, Phillips, R and Green, J. Cost implications of sexual dysfunction: the female picture. International Journal of Impotence Research. 2004 Apr;16(2):130–4.

Felix, WR, Sigel, B and Popky, GL. Doppler ultrasound in the diagnosis of peripheral vascular disease. Seminars Roentgenology 1975;10:315–22.

Sutanto, SA, Tan, M, Onida, S et al. A systematic review on isolated coil embolization for pelvic venous reflux. Journal of Vascular Surgery: Venous and Lymphatic Disorders. 2021 Aug 4:S2213–333.

Penises small, absent and disappearing

In early 2000 I gave a brief lecture to my unit describing men who had problems with their penile appearance.

Robbie was a 20-year-old student who had been worried about the length of his penis for a number of years. He was ashamed of his penile size and had therefore never been on a date let alone had sex with a partner. At interview Robbie was obviously anxious. He had previously had psychological treatment for social anxiety. He was not on any medication. On further enquiry, he told me that he had measured his flaccid penile length a number of times, and estimated that it was about 4.5 cm. He was unsure of the length of his erect penis as each time he tried to measure it, his erection faded, although he could maintain it at self stimulation for pleasure, which he enjoyed and was not problematic. Examination of his genitals indeed showed a small penis with normal sized testes, although these were somewhat retracted in the scrotum, likely secondary to anxiety. He had a body mass index of 28 (raised) and had an obvious pubic fat pad. I used a ruler to measure from the pushed back fat near his pubic bone to the tip of the glans penis on the dorsal (upper) aspect. The penile length measured 5.2 cm flaccid. That indeed is well under the average. However anxiety can cause blood vessel narrowing and consequent penile shortening. He was very keen that I should obtain an estimate of his erect penis. We discussed penile injections, and after some consideration he returned on another day. 15 minutes after injection his penile length was 13.5 cm, well within the normal range. I also undertook some hormone tests which were normal, as was his general physical examination. He was tremendously reassured by his normal erect penile length and when I saw him six months later had started a sexual relationship.

Sebastian was a 60-year-old retired stockbroker who had type 2 diabetes. He walked into my consulting room preceded by his very large abdomen. I listened to his story of erectile dysfunction that had marred his marriage for a number of years. His wife was keener to recommence sex than he was. As per routine I asked him to get undressed so I could examine his genitals. I was taken aback! His pubic area was an extension of his abdominal fat. However no penis was visible, but rather a vertical slit which looked to the world like a vulva. I think he saw my consternation and reassured me that there was a penis buried beneath the slit. After some digging and delving a found what looked like a very small penis. I discussed his case with my andrology colleagues who informed me that surgery could be undertaken to remove excess fat and skin in the pubic area. However they wanted him to lose weight before doing so. I referred him back to his GP to be seen by a local dietician. However he did not reattend.

A recent review of nearly 5,000 urology patients in Italy involved the subjects measuring their flaccid and erect penile dimensions. This was done from the pushed back pubic fat to the tip of the glans as described above. The mean flaccid length was 9.47 cm (+- 2.69) and 16.8 cm (+-2.55) when erect. A previous paper from China looking at 5,000 men entering a 'urology counselling service' put these figures at 6.5 cm (+-0.7) and 12.9 cm (+-1.3) respectively. There was some variation in these dimensions for the different Chinese ethnicities. Other studies from the USA, Israel and India give figures rather in keeping with the Chinese data.

A micropenis is defined as one less than 4 cm in the flaccid state. There are a number of causes for this.

Acquired adult buried penises usually result from obesity, but can also be the result of lymphoedema (swelling) of the area or an inflammatory skin condition lichen sclerosis. It is important to view the penis and to encourage weight loss with surgery if necessary, so as to prevent skin cancers developing in the area.

Koro is a condition mainly seen in the Far East where the patient perceives the penis to be shrinking, which induces a panic like reaction and an impending fear of danger or death. It can last for hours. The patient may attach string to the penis and pull on it to prevent it being totally retracted and disappearing. It may be part of a psychotic illness but more usually is underpinned by hypochondriasis and fear. I have never seen a case of Koro.

Chowdhury, AN and Brahma, A. Update on Koro research methodology. Indian Journal of Psychiatry. 2020 Jan-Feb;62(1):102–4.

Chen, XB, Li, RX, Yang, HN and Dai, JC. A comprehensive, prospective study of penile dimensions in Chinese men of multiple ethnicities. Int J Impot Res. 2014 Sep-Oct;26(5):172–6.

Di Mauro, M, Tonioni, C, Cocci, A et al. Penile length and circumference dimensions: A large study in young Italian men. Andrologia. 2021 Jul;53(6):e14053. doi: 10.1111/and.14053. Epub 2021 Mar 21.

Cohen, PR. Adult Acquired Buried Penis: A Hidden Problem in Obese Men. Cureus. 2021 Feb 1;13(2):e13067. doi: 10.7759/.

Post orgasm illness syndrome (POIS)

Dr Jane Ashby, and I presented two cases of POIS at a local meeting of the Jane Wadsworth Clinic in 2008. The following year she presented them to the annual SSTAR meeting in Boston Massachusetts and finally wrote them up in an academic journal. Jane was a junior doctor in training at the time.

POIS was first described in 2002 by none other than Marcel Waldinger. It is a rare collection of symptoms, mainly reported in men, following orgasm which spontaneously resolve in a few days. The major symptoms of POIS are severe fatigue, muscle pain, a 'flu like state and inability to think clearly, which patients often describe as 'brain fog'. POIS can truly devastate people's lives by causing severe mental and physical symptoms, and preventing a satisfactory sexual relationship in those who find that sexual abstinence is the only answer the problem. The cause of POIS is unknown, although Waldinger suggested allergy to the patient's own semen was possible. There are no satisfactory treatments.

The first case I saw was a 57-year-old gay man who had lived in the UK all his life and who worked as a travel agent. He had noticed a gradual onset of physical and mental symptoms that came on about two hours after orgasm. These consisted of muscle pains in his legs, 'flu-like symptoms, poor concentration and mental dullness that lasted for about three days, after which he returned to normal. He found that if he engaged in sexual activity but did not orgasm, he did not develop symptoms. I carefully examined him physically and found no abnormality. There was no evidence that he was either anxious or depressed. A wide battery of physical investigations were normal. My psychologist colleague undertook neuropsychiatric tests to see if there was a correlation between his reported mental impairment and neuropsychological scores. When comparing tests pre orgasm with those done a day after orgasm there was no significant difference. I asked my neurologist colleague for an opinion. He could not find a specific cause but said he likely had a variant of chronic fatigue syndrome. As no cause for his symptoms was found I suggested managing his symptoms by asking him to take a non steroidal anti-inflammatory drug, diclofenac, in particular for his muscle pain, just prior to orgasm. By so doing he reported an improvement in his POIS symptoms by up to 25 per cent.

The second case I saw was a 25-year-old Pakistani man who complained initially of rapid ejaculation as well as fatigue, malaise and intense lower leg pain which came on 30 minutes after ejaculation and lasted three days. He did not have any other physical or psychiatric illness. His first sexual experiences were aged 12 in Pakistan where he had active and passive anal sex with his male classmate. This continued well into his teenage years. He also suffered from very fast ejaculation. Aged 18 he had his first sexual experience with a woman but was unable to achieve vaginal penetration because of the fast ejaculation and erectile dysfunction. Over the year before he came to see me he had regular sex with a woman who was very supportive of him. When he was able to penetrate her vaginally and orgasm inside her, he had no post orgasm symptoms. The patient himself attributed his POIS symptoms to 'misuse' of his sexual organs as a teenager. Again a full battery of investigations as well as neuropsychiatric tests revealed no abnormality. Diclofenac in his case was not helpful.

Ashby, J and Goldmeier, D. Postorgasm illness syndrome – a spectrum of illnesses. Journal of Sexual Medicine. 2010 May;7(5):1976–81.

Reisman, Y. Clinical experience with post-orgasmic illness syndrome (POIS) patients-characteristics and possible treatment modality. International Journal of Impotence Research. 2021 Jul;33(5):556–62.

Sonkodi, B, Kopa, Z and Nyirády, P. Post Orgasmic Illness Syndrome (POIS) and Delayed Onset Muscle Soreness (DOMS): Do They Have Anything in Common? Cells. 2021 Jul 23;10(8):1867.

Bolanos, J and Morgentaler, A. Successful treatment of Post-orgasmic illness syndrome with human chorionic gonadotropin. Urology Case Reports. 2019 Nov 22;29:101078.

Strashny, A. First assessment of the validity of the only diagnostic criteria for postorgasmic illness syndrome (POIS). International Journal of Impotence Research. 2019 Sep;31(5):369–73.

Chapter 23

ENDINGS AND NEW BEGINNINGS

As the 2000s progressed, I felt I had more mental space to be involved in themes apart from work. I rekindled my interest in trains with a visit to the National Railway Museum at York with Lydia, a nurse colleague. The whole days excursion began just before 9 am at King's Cross station on a cold winter's day. With a diminuendo clattering, a diesel engine gently brought the old burgundy red British Rail passenger stock to a halt at the platform. Our carriage was bitterly cold as the steam heating had yet to be connected. Mayflower, a gleaming green LNER Thompson class B1 steam engine, built the year I was born, inched onto its eight carriages. Steam engines are necessarily noisy, but on the plus side they visibly exude energy. She sat there, like a powerful race horse champing at the bit. Smoke and steam curled up from the blastpipe chimney. At the front of the engine, a shrill, ear piercing, squall of boiling water burst forth from the blow down valves' copper pipes, ensuring that no water deposits lingered to choke the water tank. At the very working heart of the beast are the cylinders, where the high pressure steam is converted into moving the pistons, which are connected to the engine's giant wheels via main and accessory connecting rods. Just before a steam engine moves off the driver will open the drain cocks to allow any condensed water to be eliminated from the cylinders – again producing a very loud blast and a cloud of steam and boiling water at the front end of the engine. All very visible and audible to the keen-eyed observer.

Tractive effort is how much power an engine is able to generate in pounds force or Lbf. For the Mayflower it is just shy of 27,000. For Mallard, a type A4 engine that holds the world steam speed record it is 35,000. The Coronation Scot is 40,000. By comparison the Eurostar generates 92,000. One might say that for all Mayflower's appearance of power, she is bottom of the league. But that is to deny the delight, romance and nostalgia that it produces in aficionados.

As I stood looking at Mayflower, memories were rekindled of childhood night-time steam train arrivals at Waterloo, walking past a grimy Merchant Navy 'spamcam' that had pulled our train from Bournemouth, looking into the driver's cab to see intense orange light from the open fire grate reflected onto the black smudged faces of the driver and fireman. In later years I made some attempts to visually reproduce the romance of steam engines.

The steam era ended in 1968. However steam is very much still alive on many heritage lines and on excursion runs on the mainline system. Many working engines are parked in their sheds, waiting to be fired up.

There is the great engine's fire grate
Silent, cold, heavy, dark,
It waits for flame, an oil soaked rag
But tomorrow it will be lifeless

The fireman and the driver
Sleep tight tonight, they are not restless
For the next day to break
For the blasting ecstatic steam

And she will stay paintwork spotless and shiny
No soot, no oil, grease or coal
But soundless in the engine works
Waiting for recall and rebirth

And here I am on this non event eve
Thinking about the frosty start
On an early spring day with narcissus array
Tomorrow
That was to be, but never came.

But the steam day will come
I hope I will be there.
Today I dream of
What might have been.

However, even Virgin's leaning Pendolino electric stock can give a relaxing ride from Manchester when thoughts can roam.

Nighttime Virgin's backward train
To Euston bound while I obtain
An image of myself in glass
As street lights come and go and pass

Reflections through the window pane
Of me that odd old man again
I wish I still had frontal hair
So close your eyes, it's rude to stare!

Better that my roaming gaze
Checks out the many lights ablaze
Along the length of carriage E
Seats blue and white and orangey

I wish that man won't pick his nose
How long to London d'you suppose?
Why can't the baby stop that crying
It is so very much annoying.

Rattling Virgin supper trollies
Up and down the aisles they volley
'Do you want a cup of tea?'
'Yes please' says woman sleepily

It's Milton Keynes has come and gone
And journeys end for all and one
So farewell Pendolino train
Till you and I can meet again.

On a typical weekday I woke just before 4 am. A quick cup of strong decaf tea (sorry if this appears to be a contradiction or oxymoron, but it did look and taste strong), then 30 minutes of meditation, sitting upright on a dining room chair next to a radiator. Focusing on the breath. Just at the point where the inbreath ceases and the outbreath commences. Just there. No airflow. But I was neither a guru, nor enlightened and so distractions galore tumble out and over the breath. Anything and everything. The next meeting at work, Tricia's bad back, one of the children's college problems. I noticed them and let them pass. Sometimes I changed my focus from the breath to scan the body. Maybe I was too keen to get to the end of the half hour. Yes – I think I was. Nevertheless I was sometimes able to become quite focused, whether on the breath, the body or so-called distractions, just seeing the latter once they had bubbled up. I hoped that if ever I became clinically anxious or depressed again, the years of accumulated mindfulness practice would cushion the severity of the illness.

Then a short car drive to Queensbury Underground station, and a few more minutes meditation there.

Down Turner Road at half past four
A place to park – of that I'm sure
To Queensbury Station – gates are closed
Are all asleep, are all reposed?

I sit with engine ticking gently
I'm in the moment – sounds a plenty
And suddenly I've left the station
Moment to moment in meditation

It's five to five, switch radio on
To hear Steve Allen have some fun

Of CHAVS and CHUGGERS, small talent stars
To news at five – then leave my car.

Ahead of me at Queensbury roundabout, is the painted metal Underground roundel displayed on a pole in the middle of the circus. Not sure exactly what it is or what to call it. Queensbury station forecourt is a swirling tornado of discarded cartons and empty soft drink bottles (get fined for drinking alcohol here, you do!). The *Metro* newspaper van with its man and trolley has disgorged his plastic-bound load into the station vestibule. In winter time, we brave early morning travellers look hopefully for the station attendant to open the gates early, so we can dive into the warmth of the waiting room atop the stairs. The attendant eventually opens her office door. Was she asleep? She is one and many – smiling, serious, talkative, authoritative – and they, in their singularity, open the gates to the trains and the new day. Climb the just washed stairs to the southbound platform – where the seasons and wildlife greet me. Salt, sand and snow on the winter platform. Magpies and ravens on the live rails and pigeons cooing on the lamps lighting the precast concrete platform from overhead. Queensbury's summer dawn brings a daybreak of orange, purple and blue grey. The whiny, groany buses can unpick that tapestry of light as they leave the roundabout – but tranquillity is soon restored. With this sky's embroidery opens new hope, contentment, just for now, just in this moment.

My fellow passengers came from a wide variety of cultures and spoke sundry languages to each other when not asleep or reading free newspapers. I made it my business to say good morning to the train drivers who boarded at Queensbury who were on their way to the Neasden train depot. They always replied with a smile and nod.

I changed trains at Wembley Park onto the Metropolitan line, which took me into the West End of London at Baker Street with only one intermediate stop. From there I either walked to St Mary's Hospital or took the rather shabby, smelly and noisy Bakerloo line, or the Hammersmith and City line both of which stopped at Paddington.

The Metropolitan line is unusual in many ways.

EARLY MORNING ON THE METROPOLITAN

Is the half past five on time
To Baker Street with Auden's rhyme?
Where once rose Watkin's Tower folly
In Wembley Park of former glory

When will the 'S' stock round the corner
From Preston Road and appear to order
White lights blazing front of the train
No whistle sound but whine refrain

Dream of electric 'Sarah Siddons'
Condenser steam the heart it quickens
Memory reverie time gone by
Now Bombardier stock, kind on the eye.

Passengers climb aboard the train
Of every religion, culture and strain
Indians plenty, lots Hindu
Christians and Muslims and sometimes a Jew,
Catholics, Poles and some from the East
People not woken, faces still creased
And cleaners and makers of morning coffee
The chatty, the quiet, the dozing, the bossy.

'side Neasden Depot our train stops clean
'Signal's red, but soon be green'

Jubilee train is on our track
Going north to Stanmore back

Passengers breathe a sigh of relief,
Back to the iPhone, back to their sleep
Mascara goes on lady's face
While 'S' stock gathers speed apace.

Neasden points go clickedy clack
We're going fast along the track
Except on days we stop unscheduled
To let the drivers off to the Depot.

Air conditioned, clean and bright
'S' stock finds the morning light
You can't help but be real impressed
She's LU's only real express

It's Finchley Road and Aldgate too
Passengers varied but now no view
Glimpses of sunlight here and there,
Grand Union canal in morning air.

Journey ends at Aldgate Station
A good job done, but no elation
Maybe just recall with tear
The Metro that was of yesteryear.

The 'S' stock (made by Bombardier) is TFL's latest stock. It is modern and air conditioned.

Creative Common Attribution Generic Licence (Mattbuck)

By the end of the line at Aldgate, all have been disgorged to their morning destinations. This is how it always was on the Metropolitan line – whether steam trained (which had special condensers to cut down on the smelly smoke in tunnels),

Steam engines with condenser used on Metropolitan line

or electric engines such as 'Sarah Siddons' pulling wooden coaches,

Sarah Siddons Electric locomotive used on the Metropolitan line

Creative Common Attribution Generic Licence (Oxyman)

'A' Stock on the Metropolitan line

Creative Common Attribution Share A Likeness Licence (Dr Neil Clifton) or more up to date electric 'A' stock, that was withdrawn in 2012. (Courtesy London Transport Museum)

There was an attempt in 1896 to build a 'UK Eiffel Tower' at Wembley Park – which failed and never was completed. Sir Edward Watkin attempted to build it – as it failed it was called 'Watkin's Folly'.

In *the 1930s WH Auden wrote a poem for the Post Office about the Night Mail train to Scotland which became a classic and was written in railway rhythm rhyme. I have attempted to use such rhythm in my poem.*

Wembley Park to Baker Street can be considered an 'express line' as it misses out most of the stations that the Jubilee line stops at between Wembley Park and Baker Street.

Just before Baker Street the line has to cross the Grand Union Canal. It traverses it over a short bridge, but has to emerge from the tunnel in order to do this.

I could exit Paddington station via the Hammersmith and City line stop, walk the length of the mainline station, passing the electric Heathrow Express train waiting to depart at just after 6 am. On another platform the Penzance to Paddington overnight sleeper rested, the Class 57 diesel that had brought it from the seashore at Penzance to grimy Paddington in just over eight hours gently ticking over to provide power to the coaches that housed the still sleeping passengers. Alternatively I could exit via the Bakerloo line station that used to surface just beyond the mainline station. That entrance has since closed, to be replaced by a spanking new exit shared with the Elizabeth line. Whatever exit I used I would then walk past the large, red brick, multistorey Post Office buildings which housed the western end of the Royal Mail underground train, that ran to the central sorting office at Mount Pleasant and then on to Whitechapel. All the Post Office buildings have now been demolished. However the Mail Train still runs as a museum ride. Onwards I strode, past men sleeping rough in uriniferous smelling alcoves, blanketed under cardboard covers, dressed in grimy clothes, and onto the back gates of the Jefferiss Wing which doubled up as its fire escape. Up the metal staircases, clanging to my second floor room inside a large portacabin, which gently swayed in strong winds. It was shockingly cold in winter and sweltering in the summer. But it was my pad. I made it homely over the years. One wall of the room was adorned by a beautiful tapestry made for me by a grateful trainee doctor, Claudia Krause who is now a consultant in Cambridge. That wall hanging remains there to this day, a colourful monument to the productive work that a trainer and trainee can accomplish when both are motivated to succeed.

Once I was settled in, I opened up my computer and plunged straight into preparing for the upcoming clinics. This meant finding the paper patient notes in former years, and latterly bringing up patient details on the computer when we went paperless. Each person's problem needed to be thought about carefully. Sometimes this meant looking up research papers. Results of blood tests and radiological tests needed to be uploaded, thought about and acted on. Breakfast was about 7.30 am. People tell me that Greek yoghurt mixed in with tinned tuna is an odd combination. To my palate the combination was just right. As I spooned in the mixture, I watched a few minutes of a science programme on BBC iPlayer. My favourite was *The Sky at Night*.

Afternoons would be more clinics or meetings.

I wish now to return to the topic of patient examination. I was instructed on how to undertake general examinations at medical school, and genital examinations by senior colleagues at the London Hospital. I think it most likely that they in turn were shown 'how to do it' by their seniors, with not much thought given to the whys and wherefores of the practice. In the 1960s and 1970s it was considered 'bad form' to question ones elders and betters. After many years of practice, I have now constructed my own nuanced and empathic credo on how genital examinations should be undertaken.

Examination of the patient with sexual problems may be embarrassing or even distressing. However, if it is undertaken with adequate knowledge of local anatomy and physiology, and with sympathy and compassion it is likely to be diagnostically and therapeutically useful. Examination should always be at the patient's pace, with continuous monitoring of the patient for signs of distress.

The aim of the physical examination of men and women who have sexual problems is primarily to gather diagnostic information, building on the information gathered during history taking. Very often it will also yield therapeutic fruit. It almost always involves examination of the genitalia which is a routine and regular activity for the healthcare practitioner (HCP). But for the patient it may at best somewhat embarrassing, but in others it will be feared and cause great distress, pain or shame. Spending a few moments prior to the examination (better still a few minutes at the start of each day) empathising compassionately how patients might be feeling about the examination, will help make the examining HCP more a sympathetic ally rather than an invader of privacy and dignity.

The examining HCP might also wish to look at their own ideas and feelings of the genitalia. There may be cultural or religious reasons as to why they are uncomfortable about such an examination. These might be discussed with a more senior and experienced colleague. It should go without saying that the HCP has at least some understanding of the organic and psychological aspects of sexual medicine, as well as being reasonably informed about the relevant genital anatomy and physiology. The patient should be allowed to undress in privacy and with dignity. The examination should not be rushed and should be undertaken at a rate the patient can handle physically and emotionally. In patients who have a history of sexual assault or who are phobic, the examination may have to be deferred for some time. It should be done in warm and comfortable surroundings with maximal privacy. Phones (HCW and patient) should be switched off or silenced, and knocks on doors ignored. The patient should understand from the outset that they have control of what is being done and can signal at any time for the process to cease (e.g. if they are anxious or in pain). Prior to the examination the HCW should explain as clearly as possible what will be done. Some patients, particularly women, greatly appreciate seeing what is going on by means of a large handheld mirror. A good light is very important. The examination should be as meticulous and methodical as the patient will allow. As well as examining the relevant area, the HCW should also be constantly checking the patient's psychological and physical response to what is happening by looking at their facial expression and movement and lie of the hands and legs. Another important consideration is the need for an appropriate chaperone. The patient should be given a choice of the sex of the examiner, although ideally there should be continuity between the person taking the history and doing the examination. This will be particularly pertinent in women say from Muslim and orthodox Jewish backgrounds. Where the patient has limited movement e.g. arthritis, due consideration should be given when setting up the patient for examination.

Most attention will be given to the genital area, but sometimes areas distant from this may need to be examined. Examples of this are general neurological examination in women with vulval pain (with an emphasis on the pelvic and genital areas), cardiovascular system in men with suspected arteriosclerotic erectile dysfunction (e.g. looking for peripheral leg pulses and taking blood pressure), or presence of facial and genital hair in men with low sexual desire.

In 2005 I became chairperson of the British Association of Sexual Health and HIV (BASHH) special interest group (SIG) in sexual dysfunction. This took up some of my daytime and many out of hours work. The SIG secretary Karl Hollows was an invaluable colleague who did a lot of the planning for our annual 'ABC' day, which entailed a whole day of basic lectures and workshops on sexual medicine for doctors, nurses, psychologist and sex therapists. BASHH expected our SIG to make a profit from these meetings, which meant me approaching pharmaceutical companies for financial sponsorship for the meeting. In return they were given a 'stand' or table at the meeting, where they displayed their products to the delegates. Ethical sex shops and companies selling sexual accoutrements were also present. The SIG held a steering group meeting once a year where decisions were made about our upcoming programmes for the next year or two. By 2015 our SIG had a diverse membership which included doctors, nurses, clinical psychologists and an epidemiologist and physiotherapist. Central BASHH also expected us to arrange sexual dysfunction speakers for their general meetings every 2–3 years. Because we made a profit we were able to spend money bringing prestigious speakers over from Europe and the USA. Alas I do not think we will see those days again as speakers from abroad will nowadays have their lecture posted on a live video link. I particularly enjoyed travelling by train to and from meetings we organised out of London. We held two in Manchester. From 1997 to 2020 I wrote over 60 research papers. Weekday evenings and weekends could be wholly taken up by this time-consuming effort. One research paper in particular had to be altered 20 times before the reviewers and editors were happy with it!

A goodly amount of time needed to be spent preparing teaching material for undergraduate medical students as well as lectures and online learning for postgraduates. As I have already alluded to, I particularly enjoyed preparing PowerPoint slide lectures with embedded videos. One such video comes from *As Good as It Gets* where Jack Nicholson and Helen Hunt interact to demonstrate how a few well-chosen words can have a dramatic effect on initiating female sexual desire. Another such video was a clip from a 1942 educational film on the topic of venereal diseases, which was shown in UK cinemas. I used that in talks on syphilis. I obtained the film by going to the Imperial War Museum Film library where I was shown excerpts from longer films and chose the sections I wanted. In 2005 this was then downloaded onto a DVD and sent to me.

One method of medical student teaching and assessment was to give them Extended Matching Questions to answer. Here is an example you might like to have a go at. Answers later in the text.

1. *Refer to metabolic medicine*
2. *Examine the patient and reassure*
3. *Refer to local A & E*
4. *Refer to local child protection officer*
5. *Refer to gynaecology*
6. *Tell Police*
7. *Refer to local psychiatric unit*
8. *Refer to forensic psychology*
9. *Give medication and counsel patient about safe sex and condom use*
10. *None of the above*

For each clinical scenario below pick ONE of the management options above that you feel would be most suitable

A. *Mike is 15 years old and was born with HIV. He uses cocaine at parties, which makes him really want sex. He says cocaine is part of his party scene, but only uses it on Saturday nights. However he finds he can't get erections when he tries to put condoms on – so he has unprotected sex. He comes to you asking for Viagra.*

B. *A 35-year-old convicted paedophile comes to see you saying that he is trying hard to lead a normal sex life, but that sex with adults doesn't really turn him on – so he wants Viagra to help get erections.*

C. *A 15-year-old girl tells you she has had intercourse with her 15-year-old partner. She seems very sensible. She tells you the one time they had sex it hurt, but that penetration was possible.*

D. *A 16-year-old girl from a Muslim family has had sex with her boyfriend. Her mother comes with her asking for hymeneal reconstruction, as she will be getting married (to another man) soon. The girl seems in agreement to this request.*

E. *A 17-year-old gay man tells you he had receptive anal sex 24 hours ago with a man who, his friends later told him, was HIV positive. He wants post exposure prophylaxis for HIV.*

F. *A 14-year-old boy who has petit mal epilepsy comes to you to tell you his penis hurts. You find a torn frenulum. On further questioning he tells you his uncle has 'shown me how my penis works'.*

G. *A 16-year-old boy tells you his penis has grown smaller since he put on a lot of weight.*

I spent a half a day each week undertaking and supervising pharmaceutically sponsored drug trials, which included candidate drugs to increase sexual desire, and local penile creams for premature ejaculation. One particularly interesting study was to inject Botox into the bulbospongiosus muscles of men to prevent PE. That muscle is involved in the process of ejaculation. The theory behind this was that if the bulbospongiosus was partially paralysed, then it could not contract, or at least contraction would be delayed at ejaculation. Unfortunately times to ejaculation for Botox were no different to placebo.

I undertook one small study relating to complementary medicine, having been challenged by a patient to undertake it.

Clinic sessions are hard work for all the staff, the doctor or nurse seeing patient after patient for up to five hours. Even a few moments respite can be most welcome. A mouthful of cold tea, retying loose shoelaces, a mini reverie of past holidays or venturing into the waiting room, ostensibly to call the next patient. A brief visual survey of folk in the waiting room can be both entertaining and refreshing. The teenage girl in a very mini skirt sitting on her boyfriend's lap necking and giggling, his hand stroking her thighs, totally immersed in her partner's company, a pair of gay men nervously holding hands, one picking his lip, the other tugging his earring, a middle-aged man sitting soldier-like in a three piece suit writing notes on a number of pieces of A4 paper, a woman in her twenties obviously high on some substance, screaming rapidly mezzo soprano into her phone. A young mother attempting to control her three loud and restless young children. As I called a female clinic number, an expectant hush descended. Then a rustle of clothes at the far end of the waiting room. A woman stood up and followed me to my consulting room. She was attractive, with a sharp jawline and cheekbones, long greying wavy hair, dressed in an ankle length, flowing, loose fitting multicoloured dress. This was matched by long dangling ornate emerald green earrings and rings of many diverse shapes and sizes on fingers and thumbs. Her notes told me she had multiple genital warts which had not responded to locally self-applied applications of podophyllotoxin (prevents wart cells from dividing and so hopefully shrinks them) and a local immune stimulant imiquimod. We discussed the next logical treatment which was to spray the warts with liquid nitrogen. As the clinic was not busy that day, I asked her more about her occupation. She told me she was a 'healer' and that she used Reiki which she went on to inform me, was a form of 'energy healing' from 'universal energy' which passes through the healer into the sick person.

Although the scientist in me was sceptical, I was prepared to be open minded about the efficacy of complementary therapies for sexually transmitted infections. I was, and still am, partial to the idea of one human healing another by touch or laying on of hands.

I had written a short letter to a scientific journal in 2000 commenting on a paper the year before by Ozsoy and Ernst, who systematically reviewed available evidence for randomised controlled trials on complementary therapies to date for HIV. The treatments included herbal or vitamin supplements, stress management, acupuncture and massage. None showed a 'cure'. However the stress management increased the quality of the patients' lives. Our letter, written some three years before seeing my current patient, responded to Ozsoy and Ernst paper by quoting another study from the USA where distant healers undertook to help patients sick with advanced HIV. They defined distant healing as 'a conscious dedicated act of mentation attempting to benefit another person's wellbeing at a distance'. Of ten patients having distant healing, none died over the ensuing six months compared to four of ten dying who received no such healing. Encouraged by these results they repeated this work on 40 similar patients with HIV, but who had not progressed to develop AIDS. Significantly fewer in the 'healing' group developed an AIDS defining illness over the next six months, were less ill, had fewer medical visits and were in better psychological state compared to controls. However their immune health as measured by a CD4 white cell count, which falls as HIV disease advances, was similar in both the healing and control group.

Returning to my patient I asked her why she had not used Reiki healing on herself. She replied that she had indeed asked a colleague to 'lay hands' on her, with some benefit but not a cure. I asked her if Reiki healing could be done 'at a distance', which she told me it could. We chatted further over the next few weeks and I set up a study, with ethics committee approval, to look at the effect of distant Reiki healing on genital warts which required surgical treatment, that is cutting out the warts painlessly with the aid of a local anaesthetic. The average time from planning surgical treatment to it being carried out was about six weeks. Distant Reiki would take place in those six weeks. A photo was taken of the back of each patients' head by a research nurse. Photos were used by the Reiki healers to focus their healing. An estimate of the number of warts and their size was made by another research nurse for each patient prior to the onset of Reiki. Photos, randomly chosen, were only sent from half the patients to the Reiki healers. The patients whose photos were not sent acted as controls. Reiki healing was carried out by 12 healers focusing on the pictures for 10 minutes a day. Just prior to the surgical removal the size and number of the warts was reassessed. In order to show a reasonable effect we calculated that 130 patients in all would be needed. In fact we only recruited 27 and we lost 10 patients to follow up. Of the 17 remaining 9 received Reiki and 8 did not. At final analysis there was no difference in outcome between the Reiki and control groups. My patient and I discussed the results over the phone. She remained convinced that Reiki was a powerful healing therapy. I became rather more sceptical in spite of the low study numbers.

Sicher, F, Targ, E and Moore, D 2nd and Smith, HS. A randomized double-blind study of the effect of distant healing in a population with advanced AIDS. Report of a small scale study. Western Journal of Medicine. 1998 Dec;169(6):356–63. PMID: 9866433; PMCID: PMC1305403.

Ozsoy, M and Ernst, E. How effective are complementary therapies for HIV and AIDs? – A systematic review. International Journal of STD & AIDS. 1999 Oct;10(10):629–35.

Goldmeier, D, Madden, P, Lacey, C, Legg, K, Tamm, N and Cowen, M. Complementary therapy and genital warts. Sexually Transmitted Infections. 2005 Aug;81(4):360.

I continued to exercise on a regular basis after stopping the codeine. That would be on two to three evenings a week after work and on Sundays. I was disinterested in what shoes I wore, which in retrospect was a bad mistake. Any old casual wear would do. After 35 to 40 minutes continuous running on streets I enjoyed the mild euphoria it produced. This was my own endogenous opiates giving me a high. In the hour or two after exercise my mind seems to up a notch in concentration and creativity. And it helped to reduce my overall stress levels for the rest of the day. If I ran for much over 40 minutes I suffered from the unpleasant effect of difficulty in falling asleep. By the end of the 1990s my knees felt stiff after exercise and made crackling sounds on fully bending. Until one's late 50s one can remain self-deluded that the body will never age. But the years roll on. Oxygen, which is our life-giver with each breath, also injures our cells by a process known as oxidative stress, where the oxygen atom acquires additional electrons This same process wears away joint cartilage. I later found out that my flat, archless feet exacerbated this process by putting excessive pressure on the inner aspects of my knee joints. Early in the 1990s I was walking on wooden boards at the Hampton Court flower show when I suddenly developed quite bad knee pain. As sudden as that – no pain one minute, painful walking the next. Running exacerbated it, so I changed my exercise to bike riding. I chose a mountain bike and found that I could cycle through the pain which was certainly less than that produced by running. I referred myself to an orthopaedic surgeon colleague who confirmed on X-ray that I had moderately severe osteoarthritis on that knee, which a couple of years later also involved the other knee. I took diclofenac, a powerful non steroidal painkiller every day which helped the pain a great deal. Unfortunately over a period of about ten years it resulted in moderate kidney damage. As time went on I could no longer ride the mountain bike with ease and bought a static recumbent bike which I could use relatively painlessly by virtue of there being no weight bearing of the joints involved. However I could only walk slowly because of pain and with a marked side to side waddle. I had each knee in turn washed out with the cartilage trimmed. That helped briefly. By 2016 I was in too much pain and opted to have knee replacements. My orthopaedic surgeon Dominic Spicer, a pleasant man of few words who, after taking plain X-rays of my knees, agreed there was nothing that could be done apart from total knee replacement i.e. inserting protheses. I felt strongly that I wanted both knees done at the same time. I knew I would have to revisit opiates and only wanted exposure to them once. Dominic readily agreed. I was put on the waiting list. Three weeks later I was invited to a short seminar for patients about to have knee surgery, where I learned that doing both knees at the same operation 'was not done in the NHS'. I remained silent, but felt pleased and fortunate that I had opted to have both replaced at once. After a four-month wait I was summoned to have a preoperative assessment in terms of anaesthetic risk. I worried terribly that my anxiety induced high blood pressure would make me an anaesthetic risk. It indeed was high but I discussed with the consultant anaesthetist the fact that when relaxed – or anaesthetised – it is pretty normal. I was given a date in mid December on a Thursday. My son Nicky drove me to Charing Cross Hospital where I presented myself at 7 am. I was shaking with fear as I spoke to the check-in nurse. He had an unfortunate manner of asking questions rather robotically in a monotone. I felt he had little empathy and perhaps was in the wrong profession. I was asked to undress completely and don a paper tunic tied at the back. This was obviously necessary but made me feel vulnerable and somewhat dehumanised. I walked into the anaesthetic room and was greeted by a very pleasant young woman anaesthetist who immediately saw how anxious I was and offered to give me some midazolam, once the intravenous line was in. Midazolam is a benzodiazepine like Valium and is normally used to heavily sedate people before surgery. The midazolam worked quicky and I became happy and relaxed – what a wonderful medication! However it causes amnesia so that I don't remember anything after that. The surgery might appear mutilating to those who do not understand the procedure. The joint is opened up and the ends of both bones involved – the femur and tibia – are shaved off and are replaced with metal or plastic. I woke in the recovery room, with searing pain in both knees which appeared very swollen over the bandaging. Once fully awake I felt overwhelmingly nervous. Looking back with the proverbial retrospectroscope I believe that the strong opiate (oxycodone) they had given me at the end of the operation and continued to give me for another week had caused this huge spike in anxiety. So revisiting my drug of addiction turned out to be a nightmare rather than a euphoric experience. And maybe that was a good thing! I found it hard to sleep because of the pain and anxiety, added to which my legs were attached to automatic pneumatic compression devices which were noisy and intermittently compressed my legs to prevent blood clots. Some of my colleagues visited me the evening after my operation, when I think I was confused about who they were. On the weekend my family came to visit. Because I had been given a spinal anaesthetic as well as the general anaesthetic I had a urinary catheter indwelling for two days, after which I hobbled using a Zimmer frame to the toilet. In spite of daily constipation ward rounds, where laxatives were offered to all patients my bowels refused to work for ten days. The physiotherapist who came the day after my operation was one I had worked with in the past in my department. Leg movements were at first small and the pain large, but I practised all day long, wanting to leave the ward as soon as I could. One of my fellow patients was a young soldier who worked in the Household Cavalry. His horse had bolted and threw him, resulting in a badly broken shoulder.

Extended matching question answers
A9, B8, C2, D5, E3, F4, G2
By day five I could walk reasonably well with crutches. I was assessed by the occupational therapist and discharged at day six. I continued to take dihydrocodeine (a less powerful opiate than oxycodone) for pain for six weeks. During the first two weeks I continued to inject myself with an anticoagulant – used to stop the development of a blood clot. At week one after the operation the wound on one leg looked a bit yellow – so I was given antibiotics by my GP. I continued exercises daily under supervision of a private physiotherapist and was discharged from the care of the orthopaedic department after eight weeks. The swelling subsided over three months. Walking was now painless. After ten weeks I stopped all analgesics and after a rather uncomfortable opiate withdrawal period of a further six weeks I was ready to go back to work. I continue to exercise on my recumbent bike and walk but do no high impact sports to save my precious new knee joints.

A few words about three people who I worked with at St Mary's.

Daniel Richardson was a trainee registrar who was with us at St Mary's for four years. Rather than doing research in HIV, which at the time was very much the fashion in the department, he opted to concentrate on sexual medicine. He was a great help to me for those years, co-authoring many research papers and having many good ideas that we both put to the test. It only took him a matter of a few months before he was lecturing at international sexual dysfunction meetings and indeed chairing sessions. His move to Brighton in 2004, where the majority of his work is in HIV and STDs, meant I no longer worked so closely with him. However, he continues to write on the topic of syphilis, which is an interest common to both of us and is gracious enough to ask me to co-author with him from time to time. He has recently been given a professorship by University of Sussex.

I believe I met Ali Mears at a national conference in Brighton in 2003 (she disagrees where we first met). She was finishing her training in GU Medicine at St Thomas' Hospital and I knew would be looking for a consultancy in due course. I was impressed by her knowledge and enthusiasm as well as her interest in sexual medicine. After that meeting I suppose I headhunted her and strongly advised her to apply for the post of consultant that had come up at St Mary's. She had already done an introductory course in sexual medicine (Institute of Psychosexual Medicine). I was not on the interview panel. However her presentation and credentials in HIV and GU Medicine were also excellent and her enthusiasm and pleasant personality gave her the job. Ali has been a unique and special colleague through the thick and thin of things over the last 16 years. I saw her as my 'work wife'.

Special mention must go to Agnes Kocsis, my psychologist colleague who I have worked with in clinic on one whole day each week for the past 20 years. Our combined clinical effectiveness was only matched by the fun we had on those Tuesdays!

I would have liked to tell myself and you that all was well till I caught Covid. However, I became increasingly worried about my health in my late 60s. Visits to the high blood pressure clinic and blood tests taken then filled me with dread. In youth and middle age, health anxiety scares would be many years apart. I saw a clinical psychologist just prior to Covid to discuss my situation. I found her to be unhelpful, with her saying I should do breathing exercises to help anxiety. Fair enough but she didn't really specify quite what should be done with the breath.

In January 2020 the world began being concerned about what would later be termed COVID-19. Three months before that my wife and I had visited our youngest daughter in Israel. I was as carefree as a worrier could be, enjoying going to the gym there as well as taking my very young granddaughter in her buggy along the banks of the Yarkon River. They visited us in the UK in February 2020, when the viral infection from China was growing into more than a whisper. By March I had stopped going to work on the Underground, parking my car in a hotel car park in Paddington. I saw my last face to face patient at the end of that month. No one was wearing facemasks at that time. I became ill ten days later with a cough, mild fever and severe malaise. My fear remained high and continuous. In less than ten days it had evolved into a combination of anxiety and depression. A psychiatrist colleague at work suggested to my GP that I should go onto a small dose of antidepressants. It did not help. Another feature of my Covid infection was losing my sense of smell for about eight weeks, finally agreeing with my wife that the nasty smell under the fridge was not her imagination but rather rotting food! After about eight weeks I returned to work remotely from home but found the video consultations stressful in that one often lost the link to the patient many times during a consultation. However I soldiered on till August 2020. My days began at 1.30 am waking with multiple panic attacks. By the middle of the afternoon I felt better and could go for walks in the nearby green belt, even deriving momentary pleasure from seeing a sunny yellow rapeseed crops. However I knew the horror of what the next morning would bring. I saw a private psychiatrist who prescribed other medications which were unhelpful. I went on long-term sick leave in August 2020. By autumn I had transferred my care to an NHS psychiatrist. I was keen to have a course of ECT as it had helped 25 years before. I began my ECT course in mid March 2021. It did not help. I was still waking terrified early each day. By Spring 2021 I had decided I could not go back to work so went for permanent retirement (I had retired and was in receipt of an NHS pension some seven years before and had returned to work on an annually renewed contract). I developed a heart rhythm abnormality in May and had to go onto anticoagulants and stay on them for life. This makes my blood thinner so I am not so likely to develop a brain clot but more likely to develop a cerebral bleed. My health anxiety of course peaked again.

At the end of 2020 my psychiatric state was monitored in the NHS by Dr Amy Jebreel. She listened and was both reassuring and attentive. I started a third antidepressant tablet in June 2021, very gradually increasing the dose from a quarter of a pill to two pills a day over eight weeks. The early days of taking that were very difficult, with exacerbation of the anxiety and frequent nightmares. Going through hell to get to improvement is a tough route to follow. But finally things did get better. Autogenic training also helped my recovery. This consists of silently repeating formulae such as 'My arms are heavy and warm', while 'passively focusing' on that area of the body. I tried to return to mindfulness briefly, but that only seemed to exacerbate the depressive state. My hope that it would buffer any overt psychiatric illness proved false.

So here I am, almost 30 months after severe Covid finally, having some semblance of a normal life. Did Covid cause an encephalitis i.e. was my brain physically affected by the virus or was it all a horrible psychological response to a highly emotionally traumatic situation? I am not sure. I am still anxious in the mornings but can sleep till much later on. I find myself no longer working as an NHS employee after 50 years and no longer working at St Mary's after 40 years. Retirement indeed feels like a foreign land. The past contours of my life have changed, and I do not yet know the new topography. I must learn anew from scratch how to fill my days. I cannot say I 'enjoy' retirement – I would say there are rare moments of pleasure. I cram my day full of writing this book (what will I do when I have finished?), learning a foreign language, teaching myself to play the piano and doing an online astronomy course. I wrote a short article about how my stammer affected my life – for 'STAMMA' the national charity for stammerers. My memory for names is awful – it always was, but possibly the ECT has made it worse. So learning the new language is much harder than the astronomy which makes logical sense.

Days that in the past were crammed full of the practice of medicine now contain large gaps where self-criticism can easily loom large. Would that my maths were better. I then could have become an astrophysicist. But they have to give lectures and seminars too. So back to the stammer. I can't seem to dodge it. Maybe if my mild ADHD had been addressed early on the stammer would have dissolved away? I recall all the treatments I have tried for rid me of this oral burden. Relaxation therapy, mindfulness, personal construct therapy, public speaking with earphones so that my voice is played back to me with a small delay, speaking slowly, paradoxical stammering where I stutter on purpose, the Alexander Technique ... the list goes on and on. And what helped in the end was accepting, at least to a degree, that I am very far from being an orator. Knowing that being tired makes it worse and that exercise and tea, taken in a timely manner, can help in set piece situations. Colleagues feeding back to me that 'your stutter is better than it used to be' only served to make me feel I was much 'less than' then and that improving slightly on a disaster is still a catastrophe. Even Tricia, my own life partner, could not help bursting out laughing when we were both in our teens and I attempted to purchase a box of 'Hamlet ' cigars (only ever smoked five) and the words would not come out – just 'Ha ha ha'. She did not laugh out of ignorance, or because she thought I was stupid, but merely because there I was trying to push the word out, with the shopkeeper wondering what on earth I wanted. After she nearly collapsed on the floor with laughter she gave me a hug and told me she loved me all the more for having a stammer. But it hurt. It always hurt. I could have given up my medical career as a student, and twice when I was ill with anxiety and depression caused by my stammer. But I stayed the course. Well done David! I lost three battles but in the end won the war – just by gritting my teeth, ploughing on and naked tenacity. I feel that perhaps I was my most critical judge of my speech. Others just accepted that stammering was just part of David Goldmeier. Maybe there is more general acceptance of speech dysfluency nowadays. That is the wish of STAMMA. But a person with a stammer will never be a News announcer or give a TV weather forecast. There are no speech Paralympics that I know of. We stammerers must therefore reach for the sky and make career and relationship choices despite our speech defect. OK, there are some professions we can never aspire to – actors, presidents, prime ministers. But mostly the world is our oyster – we just need to be tough from start to end and reach for our star.

I suppose retirement would have come even if I had not suffered Covid. Maybe I should have planned better?

I have reached the end of my story. I am staying with my son and his family in north Manchester. I take a walk in Heaton Park and stroll into its pleasure gardens. I pick up pace and walk briskly with the mid afternoon sun on my shoulder and the majestic Pennine Hills ahead of me. I take a wooded path heavy with shrubbery on either side and canopied by sturdy oak and ash trees. I reach the slightly crumbling redbrick walls of what was once a kitchen garden, maybe 200 years ago. Beside it, behind further greenery is a large boarded cottage that I guess must have been a bothy for the garden staff employed by the owners of the great house that lies over the hill. There are no records of the gardeners and journeymen who lived in the cottage. In spite of this I sense what they left behind. I feel their soil bound lives would have made them contented souls. As I gaze on the scene the evening light spreads a blanket of contentment over me.

Maybe I didn't do too badly after all.